CURE
DIABETES
PARKINSON'S
&CHRONIC DISEASE

A New, Definitive Cure for Many Chronic Diseases. Medical Fallacies Exposed. Why Modern Medicine is Wrong, & Your Doctor is Clueless. How to Save Your Life.

ROBERT S. FARMER, M.D.

ISBN: 978-1-4834-7476-2 (sc)
ISBN: 978-1-4834-7475-5 (e)

Lulu Publishing Services rev. date: 10/10/2017

The reasonable man adapts himself to the world; the unreasonable one persists in trying to adapt the world to himself. Therefore all progress depends on the unreasonable man. — Nobel laureate George Bernard Shaw

All truth passes through three stages. First, it is ridiculed. Second, it is violently opposed. Third, it is accepted as being self-evident. — Arthur Schopenhauer

It is always a sign of mediocrity in people when they herd together ... The truth is sought only by individuals, and they break with those who do not love it enough. — Dr. Zhivago (by Nobel laureate Boris Pasternak)

Medicine likes to think it is the most 'beneficent' profession, but it is deeds not words that count. — Roy Porter

It is part of the cure to wish to be cured. — Lucius Annaeus Seneca (Seneca the Younger, ca. A.D. 50)

You must wager. There is no choice. You are already committed. Which will you choose, then? Let us see. — Blaise Pascal

Dedication

To my mother Lois for always giving me the freedom to be myself. She always let me go down to the creek to play, even though she knew that I would always return soaking wet.

Also to Sweetie, the love of my life.

Acknowledgements

Thanks to Dr. Jerroll Dolphin of St Luke School of Medicine and Dr. Orion Tulp of the University of Sciences, Arts, and Technology (USAT). You guys made it all possible.

Thanks to Doug Koschalk for printing help and some other more general kinds of help.

Thanks to Sara Minor for helping me with some computer stuff.

Thanks to Ralph and Kay David for other miscellaneous life help.

Thanks to Doy and Sandy Prater for helping me with miscellaneous stuff when they really didn't have to.

Thanks to Dad for financing medical school, etc.

Thanks to Valerie Raba.

Contents

Introduction

Cur'd yesterday of my Disease, I died last night of my Physician. — Matthew Prior (1714)

Why buy this book? Well, <u>the first reason</u> is that this is the most innovative and important medical book that has ever been written—I've accomplished cures that no other doctor has ever accomplished—already 50+ diseases, and probably more than 100+ diseases eventually. You cannot find this information anywhere else on earth. I've discovered that nearly everything that you think of as old age etc. is a treatable, chronic infection and is therefore largely reversible.

<u>The second reason</u> to buy this book is that everyone, including you and your family members, has or will have chronic health problems, perhaps debilitating chronic pain, and <u>doctors currently don't have a clue</u> how to treat them—so what happens is that doctors try truly insane treatments that are based on outdated, defective, biased, profit-driven research, and those treatments are often useless and harmful, not to mention fabulously expensive. You are spending way too much of your hard-earned money on health care, and that health care often doesn't work—it often makes you sicker (but you might not know, because you're not a doctor)—and that increases the cost of health care even more. Pretty soon you have to choose between buying food or some overpriced medicine that probably doesn't work anyway, but you're not a doctor, and your doctor (whom perhaps you love and/or trust) tells you that you need it, so are you going to spend a hundred dollars a pill or whatever, or are you going to eat? No one should be put in this position. Modern medicine is a scam—or at least much of it is a scam. Doctors are ignorant, clueless, and far too arrogant to consider

the possibility that their methods don't work, at best, and often kill people, at worst. For instance, the new generation of immune-system-destroying drugs that you see advertised on TV is the new version of blood-letting—a genocidal scam that should be outlawed.

The third reason to buy this book is that you are spending far, far too much money on medical care, whereas my approach to medicine and health is cheap—dirt cheap, especially compared to what the American Medical Association would like to sell you. I'm talking forty cents or less per treatment, and generally less than five dollars a day overall, whereas going to a hospital generally costs $2,000 just to walk in the door. Not only will you save your own life, the lives of your loved ones, and the lives of your also-loved pets, you will save a huge pile of money. You will save the cost of this book many, many times over.

Part of the reason that I became a doctor is because I thought it would be one of the most ethical ways to make a good living, but I now see that doctors have always been incompetent—incompetence that goes back for thousands of years. The difference between someone such as Hippocrates, the so-called father of medicine, and modern doctors is that Hippocrates seemed to recognize his limitations, whereas modern doctors know very little more than Hippocrates did, yet they think they know everything, despite their almost-continual failures.

I need to insert a legal disclaimer that I can't guarantee that you will cure every disease after reading my book, however, there are clearly many, many (i.e. all) people—a vast majority—who are the victims of medical incompetence (unfortunately different from medical malpractice, which is a legal term), who can benefit substantially from what I have to say, regardless of whether their diseases are actually cured completely.

I have now cured many more than 50 diseases that no one else has cured, including definitely these: autism (one case—Asperger's syndrome—100% success rate), osteoporosis, chronic pain/fibromyalgia, allergies, obesity, emaciation, kidney disease, urinary/bladder problems, heart disease, lung disease (COPD, asthma), Parkinson's Disease, multiple sclerosis, ALS, irritable bowel syndrome, celiac disease, diabetes mellitus, arthritis, eye problems (e.g. diabetic retinopathy, glaucoma, macular degeneration), plantar fasciitis (foot pain), wrinkles, tooth and gum disease, chronic fatigue syndrome, colorectal problems, gout, and hair loss. In addition, other diseases that I've probably cured include cancer and Alzheimer's disease as well as various problems of any part of your body (e.g. gynecological)—many diseases can be cured or significantly improved (e.g. cystic fibrosis, schizophrenia) using the information in this book. How is that possible? It's because these diseases are caused by one (or a few) particular infection(s) that has been passed from mother to fetus for millennia, and because of the way that it is transmitted, it avoids and outsmarts the immune system and is never recognized by the body as foreign. (Burnet and Medawar won a Nobel Prize for a related concept.) This slowly-growing infectious parasite is then mistaken for old age and many diseases, and people die without ever realizing that they had a treatable infection. Doctors adamantly refuse to consider this infection despite the fact that it, or a very similar one, is described in every medical reference book! Meanwhile, using my methods, I've improved my health to the point where I feel confident that I am healthier than any other doctor my age or older—as a group, doctors are some of the sickest-looking people whom I've seen, anyway—which kind of makes you wonder, doesn't it? I have a policy that I don't take health advice from doctors who are obviously sicker than I am. I haven't taken any other doctor's advice for quite a while, now.

Here is just one example of how clueless doctors are: If you ask a doctor about hypertension (high blood pressure) and kidney disease, he/she will say that hypertension causes kidney disease. That is exactly backwards! Hypertension doesn't cause kidney disease—kidney disease causes hypertension! (You'll have to think about that for a long time, probably, but for now, just take my word for it. The kidney disease comes first. I explain it in this book. Also, my approach to medicine will make kidney dialysis and transplants obsolete—which is why a lot of kidney surgeons and others don't want the truth to come out.) Doctors often confuse cause and effect! They often don't understand even the simplest concepts! Why? I wish I had a good answer for that, but the answer appears to be just that doctors are narrow-minded, passive to authority (institutionalized ignorance), trapped in primitive thought patterns, and incredibly arrogant. They say things such as, "That's not in the literature," as if that proves something—not understanding that the people who control the literature have huge financial motivation to suppress the truth. (Doctors never learned to think; they learned only rote memorization. If they had learned to think, they would realize that modern medicine is often insane, with no cures for many curable diseases, and many specious rationalizations.) Of course "that" is not in the literature, because medical journal editors don't want you to know that they are lying to you! Curing diseases is cheap, but managing diseases is often very, very expensive and therefore profitable for some people. If you own a medical journal, how are you going to stay in business? How is my father's doctor going to pay for his Bentley? (Or at least there was always a Bentley in the parking lot when I drove my father to his doctor's appointment.) By telling people to use more herbs and spices (cheap), or by convincing them that they need a very expensive patented drug that they need to take every day for the rest of their lives, and which needs to be advertised in a medical journal every week or month (profitable)? (By the way, my father was recently killed by his doctors—they gave him a fatal infection during hip surgery.) Doctors have killed all of my dead relatives with their incompetence.

I spoke on the phone to the editor of a medical journal entitled something like, "*Diabetes and Endocrinology*" before the first issue was printed. I said, "I've cured diabetes mellitus, and I'd like to write an article that will fit your criteria." She responded, "We're not interested." So I said, "You're not interested in curing diabetes mellitus?" And she said, "No, we're too busy." This is obviously a lie. Similarly, an editor at the *Journal of the American Medical Association* (JAMA) told me, "We are interested only in articles about patented drugs." Advertisements for patented drugs and other medical products keep the media in business. How many times have you watched the network evening news on TV without seeing at least one (and usually several) drug advertisement(s)? Answer: never! Drug companies essentially own the media outlets, so you cannot depend on them to tell you the truth. All they want is your money, and they don't seem to care if they have to commit genocide to get it. Consider the warning for Humira™, a drug that is designed to suppress your immune system:

"*Humira™ can lower your ability to fight infections, including tuberculosis. Serious, sometimes fatal infections and cancers including lymphoma have happened, as have blood, liver, and nervous system problems, serious allergic reactions, and new or worsening heart failure.*"

Does that sound like good medicine to you?! The fundamental principle for these many drugs is based on a delusion! Autoimmune diseases are not real! Modern medicine is literally insane! And you are paying for it with your money and your lives!

Consider two doctor-associated fitness experts who dropped dead in an instant: Jim Fixx, a running expert/author, and Bob Harper (from *The Biggest Loser* TV show). Jim died, but Bob was revived. If they (or their doctors) know so much about health, then why did they drop dead? Huh, smarty pants?! Well, I know the reason, and it's in this book.

In this book, I explain the diseases and the history, and I give detailed information about how to eliminate or improve your disease, using terms that everyone can understand. You've *possibly* or even *probably* wasted hundreds of thousands of dollars on confident-but-incompetent doctors; don't you think you can afford to gamble a few dollars on this book to get the truth? If their way is the only way, then why can't they cure even a simple disease such as Multiple Sclerosis? I've cured it, and it's pretty simple! Or esophagitis! Or Irritable Bowel Syndrome? Or gluten allergy? (Remember how good fresh-baked, hot-buttered bread tastes? Remember pasta?) Simple, simple, simple! They have no answers, but they nevertheless refuse to consider any other possibility—that is literally insane! Do you really want to be treated by an insane doctor? Do you?!

If you still don't believe me, read *David and Goliath* by Malcolm Gladwell, and study the chapter on Dr. Freireich. He and I are a lot alike. We both fight for the truth regardless of what clueless critics think. However, my medicine is better than his—that's just part of evolution.

During this time (2017) when there is so much talk about repealing Obamacare and instituting some Republican overhaul of health care, and whether preexisting conditions should be covered, people are missing the far-bigger point: Your doctors are clueless! They don't know what they're doing about 80% of the time! They're killing you! Who cares if you have health care when a trip to the doctor will result in a prescription for a drug that will literally kill you? Many people would literally be better off not going to a doctor! If I hadn't become a doctor, I'm sure that I'd be dead by now. I don't know how so many doctors have gotten away with so much incompetence for so long, and have made so much money doing it.

Probably 99% of diseases that are believed to be incurable really aren't.

They're just not curable by stupid, narrow-minded doctors who sold all of their medical school textbooks at the end of each semester because they were convinced that they knew everything. Just as Deep Throat told Woodward and Bernstein to "follow the money" in the Watergate scandal, in medicine, we need to "follow the chemistry," and that's what I've done. For me, I cured my Parkinson's disease, diabetes, and other related diseases, and I now look younger and am healthier than 99% (or more) of the people on earth who are my age or older. You too can *feel* younger and healthier—and *look* younger and healthier, with fewer wrinkles and more tight, smooth, firm skin. It's not that hard—you just have to understand the truth and be motivated to adjust your lifestyle a bit. You'll feel better, look better, live longer, lose your pain, and save a fortune. It all starts right here.

—Robert S. Farmer M.D.

SECTION 1

A Basic Primer

Primum non nocere. (First, do no harm.) — Hippocrates

The historical record is like the night sky: we see a few stars and group them into mythic constellations. But what is chiefly visible is the darkness. — Roy Porter, *The Greatest Benefit to Mankind*

Political revolutions, the writer Amitav Ghosh writes, often occur in the courtyards of palaces, in spaces on the cusp of power, located neither outside nor inside. Scientific revolutions, in contrast, typically occur in basements, in buried-away places removed from mainstream corridors of thought. — Siddhartha Mukherjee, *The Emperor of All Maladies*

It is in vain to speak of cures, or think of remedies, until such time as we have considered of the causes . . . cures must be imperfect, lame, and to no purpose, wherein the causes have not first been searched. — Robert Burton, *The Anatomy of Melancholy* (1621)

Life is short, the art long, opportunity fleeting, experience fallacious, judgment difficult. — Hippocratic Aphorism

Chapter 1

Everybody Gets Sick

All Truth passes through three stages: First, it is ridiculed. Second, it is violently opposed. Third, it is accepted as being self-evident. —Arthur Schoepenhauer

Everyone knows someone who is getting older and suffering the aches and pains that accompany aging. It is not uncommon to know someone with a chronic disease such as diabetes mellitus, heart disease, lung disease, kidney disease, mental illness, vision deterioration, hearing loss, or chronic aches and pains. A great many people who seem otherwise healthy have severe pain in the back, neck, or feet that seems to have no cause—the pain just seems to have come out of nowhere, and there seems to be little or nothing that can be done to treat it—it just seems to be a natural part of aging. Going to bed often becomes the treatment, but all-too-often, people feel worse when they wake up in the morning than when they went to bed the night before, such as with the morning stiffness of rheumatoid arthritis.

We have probably all heard about the person who threw his or her back out by bending over to pick up a newspaper off the ground, or to pet the family dog, or to pick up a cat or a child. We have probably all heard about how old people shrink as they grow older. These are due to severe deterioration of the body that often appears decades sooner for some people than for others. We have all heard about the fitness experts who suddenly dropped dead of a heart attack with apparently no warning, and people often think, "no one saw it coming. If it could happen to him/me, it could happen to anyone." These often-tragic events have always been largely unforeseeable, seemingly random acts of God or gods or nature or just bad luck. There was always little or nothing to be done about it—just accept the inevitable.

When I was growing up, I got the same standard health information that everyone else got, I imagine: if you eat your vegetables, get a balanced diet, drink your milk, and get regular exercise, you can live to be a hundred years old. Like most people, I believed that. I believed that the people who told us that were experts who knew what they were talking about. I followed that advice. From a young age, I tried to be careful about my health, becoming a vegetarian, drinking lots of milk, exercising fanatically—even to the point where I could do pull-ups using only one arm—and for my efforts, I suffered constant pain; a continuous, gradual deterioration; loss of energy; and very nearly an early death at the age of 49, just like my mother.

My mother had been an "alcoholic" and a smoker. Like everyone, I had been brainwashed to believe that alcohol and tobacco were harmful, and I pressured my mother severely to stop, "for her own good." She finally did stop smoking and drinking alcohol, and she died about six months later, of "cancer." Over the next few decades, my own health continued its precipitous decline, despite my best efforts to eat well and exercise with my meager financial resources. After some indefinite period of time, I couldn't help but notice that the model of health that I had been taught all my life, the model, the paradigm that I believed with near-fanaticism, wasn't working! My health was terrible! This, despite the fact that I was obsessed with making it better! And my mother had died so young—when quitting alcohol and tobacco was supposed to save her health, instead, quitting had definitely seemed to have the opposite effect, hastening her death. Her doctor's (some spoiled little punk with a solid gold Rolex watch) explanation sounded utterly stupid: "She was just too addicted to alcohol and tobacco, and her body couldn't handle quitting." If alcohol and tobacco are supposed to be so bad for people, then how could quitting their addiction kill them? That's illogical! The doctors had no answers. My own health continued to deteriorate, and after I found myself living in Baltimore, Maryland—a place with several universities—I decided to try to become a doctor, much to my father's dismay. Although he eventually helped me to get through medical school, he made it clear that he was not happy about it.

I eventually graduated with an M.D. degree, and because of political harassment, I had to transfer to another medical school and repeat many of my courses—essentially going to medical school twice—to get a second M.D., along with an M.B.B.S. degree—a British medical degree.

As a medical student, I had noticed something interesting about the people who would become doctors: after every semester, they would sell their textbooks. I wondered: do they all have such photographic memories that they can remember literally everything in those incredibly dense books, or how can they justify getting rid of so much knowledge? In contrast, I kept every textbook, because I knew how fallible my memory was, and I wanted to be the best doctor that I could possibly be, and I knew that I might want to refer to those books someday. My health was still bad, and I wanted to have the best doctor on earth treating me, and if I had to become the best doctor on earth to do that, then so be it—so I was going to need my books! Why shouldn't I become the best doctor on earth? Somebody has to do it—why not me?

Here is a list of some of the problems that I have had to deal with over the decades: severe, chronic pain, throughout my body but especially in my neck (fibromyalgia); Parkinson's disease; multiple sclerosis; constant fatigue (chronic fatigue syndrome/adrenal insufficiency); scoliosis (sideways curvature of the spine); deterioration of teeth, gums, and jawbone so that my jaws did not connect properly; tooth sensitivity to cold liquids and sweets; gray hair; wrinkles; double chin; jowls; shoulder bursitis; chronic severe stomach pains since childhood, diagnosed as a stomach ulcer; plantar fasciitis (foot pains); torticollis (tightening of the muscles on one side of my neck, pulling my head to one side); depression; anxiety; irritability; loss of concentration; eye and skin sensitivity to light, especially fluorescent lights; extremely sun-sensitive skin; celiac disease/gluten allergy; other undefined food allergies; gastrointestinal paralysis (ileus); alternating constipation and diarrhea; combined constipation and diarrhea; balance problems/vertigo; hypothyroidism; weight (fat) gain; chronic muscle degeneration (sarcopenia); excessive appetite; inability to eat and keep food down; severe gastroesophageal reflux disease (GERD); esophagitis; scotomas (spots or "floaters") in my vision; changes/deterioration of vision; loss of sense of smell; seeing flashing lights at night with my eyes closed; restless legs syndrome; carpal tunnel syndrome; chronic "tennis elbow";

"age spots" on the backs of my hands; palmar fasciitis (tightening of the palms of my hands); muscle spasm/tightening on the left side of my body, mainly; "growing pains"; degeneration/alleged fusion of the spine (probably ankylosing spondylitis); hormonal deficiencies, e.g. thyroid, etc.; long-term deficiencies of sodium and other electrolytes; heart irregularities (especially bradycardia, or slow heart beat) and weakness; Chronic Obstructive Pulmonary Disease (lung disease—emphysema/bronchitis); very slight asthma; spasms of the esophagus; diabetes mellitus, type 2; osteoporosis with possible vertebral fracture and four broken legs at the ankle spread over three occasions; lipomas (fatty tumors just under the skin); probable lymphoma; severe, widespread acne, especially of the face and buttocks; arthritis, mostly just of the back, ankles, and spine; ganglion cysts (large bumps on the wrist bones) and numbness of hands and feet.

Whew! It didn't seem like that much until I started to write it down! But I have cured or at least drastically improved all of those diseases and conditions! However, I can assure you that it has been a miserable—and completely unnecessary—journey, one that could have been prevented if just one doctor had bothered to do his or her job competently, diligently, and properly. Just one. And I see people all the time on TV and the internet who are really suffering—and dying—from a disease that was first recognized 140 years ago, and which doctors adamantly, even violently, refuse to recognize! And it is so simple! So very, very simple! At least conceptually . . . but that doesn't mean that it is easy to treat. Successful treatment takes years, maybe decades, and probably the rest of your life. Treatment requires making a new lifestyle choice, and a person has to be highly motivated and committed, but treatment is not that difficult or onerous. Life will still be enjoyable.

So why haven't you heard about how to cure these diseases before? Well, I can't say with absolute certainty, but it certainly appears that powerful people are trying to suppress the truth for their own financial gain. The other possibility is that they are just stupid; it seems that every artist, businessman, lawyer, carpenter, plumber, musician, etc. who has never so much as taken an entry-level college biology course thinks that he or she knows more about medicine than I do, with my seven college degrees, including three medical degrees. I feel a lot like Ignaz Semmelweis, who, it is said, was driven insane and died because of his inability to get doctors and nurses to wash their hands in the 1840s. Today, it goes without saying that health care workers should wash their hands, although they still don't always do it, even now. You should ask [censored] of [censored] News why he refused to report the truth when I put my articles directly into his hands. You should ask Dr. [censored] why he still refuses to report the truth when I put my articles directly into his hands. Ask why [censored] won't tell the truth. Ask why all of the TV news doctors have never returned my attempts to contact them. Ask why [censored], the head of the [censored], won't tell the truth. The most likely hypothesis is that because most of their income comes from selling advertising to drug companies who sell useless and harmful drugs, and because they have the most to lose by acknowledging the truth, they have decided that being multi-millionaires is better than being honest, or they're too ashamed to admit how stupid they've been.

I've never been a particularly good salesman. It often seems that I couldn't sell a glass of water to a man dying of thirst in the desert. I'm an extremely logical person, but I've finally come to realize that most people don't usually make decisions logically—they make decisions emotionally, and I apparently don't know how to convince people using emotions; instead, I'm a hard facts, scientific kind of guy. I like testable hypotheses, with obvious, verifiable conclusions. In the Watergate scandal that forced President Nixon to resign, Deep Throat (i.e. Mark Felt) always told reporters Woodward and Bernstein, "Follow the money." In trying to solve the mystery of chronic disease, a similar process is required, but in medicine, the mantra should be, "Follow the chemistry." The biochemistry book that I bought for medical school had approximately five previous owners before I bought it used. This means that those five doctors really don't know very much about biochemistry, because there is no way that they have remembered everything from that book. By selling their books, those doctors deliberately chose to be ignorant, and I'm positive that virtually all other doctors have made the same mistake—this is why I'm the one who cured these diseases, and not someone else—and their patients are paying the price every day for their doctors' ignorance. I kept my books so I could continue to keep learning. The very first time that I saw the chemical pathway for the catecholamine hormones, I instantly hypothesized the cause of Parkinson's disease, and, as it turns out, that was about half of the problem, right there. I have since consulted that book many, many times over the years. I have done a great deal of reading over the years, and a lot of experimentation, and because I am open-minded and extremely logical, I realized that the predominant medical beliefs—my medical beliefs—were wrong, and I had to change my thinking 180 degrees, which was a very painful process, literally too painful for most doctors, apparently. Now, though, I am quite likely the healthiest 56-year-old person on earth, thanks to chemistry. I like the certainty of consistent chemical reactions and I like finally being healthy again after all these years. I have now developed an understanding of health and medicine that is 100% logical and consistent, unlike the predominant medical model, which is filled with logical inconsistencies, contradictions, and fanciful, delusional rationalizations. This book will explain how modern medicine went off the rails, how to cure a large majority of chronic diseases, and how to stay healthy. There will be some technical explanations to satisfy doctoral-level scientists, but regular people will be able to skim over those without any problem. By the time when you finish reading this book, you will be among the first people in the history of the world to truly achieve control over your health.

Chapter 2

A Realistic Self-Assessment

Who are you going to believe—me or your own eyes? — Groucho Marx

Modern medicine is a disaster. I see many, many people on TV and in my neighborhood who are really suffering, often with extreme pain, and I see that the doctors who are treating them are completely incompetent, increasing their patients' suffering enormously and killing them slowly and at great expense, all while bankrupting them. When a 500-lb woman tells me that she is very happy with her doctor, something is wrong—there is absolutely no way that a 500-lb woman should be happy with her doctor! For many people, the cost of their often-counterproductive health insurance is their second-biggest expense. Out-of-pocket treatment is ridiculously expensive, and insurance often seems to be a rip-off, and—what many people don't seem to have noticed—medical treatment usually doesn't work for chronic diseases, except perhaps occasionally by providing short-term benefits at the expense of long-term health. There's a simple reason for that: modern medicine is usually wrong; <u>doctors have zero understanding of chronic diseases. Instead, they treat every sign and symptom separately, and they never look for an ultimate cause.</u> (—except when they look for genes, which is almost-always the wrong track. Doctors do all right with acute diseases, though—if it's a bacterial disease that can be treated with an antibiotic.) In addition, they are far too quick to jump to the always-wrong diagnosis of "autoimmune" disease, which appears to be a complete fallacy; you might as well say that diseases are caused by evil spirits, sprites, and elves. I've cured several "autoimmune" diseases, and they are not "autoimmune" at all; they are an infection that doctors are just too stupid to recognize. The simple answer is that in approximately 80% of cases, doctors have no idea what they're doing, so they use some formulated guidelines, and those guidelines are wrong, because they are just a guess, a guess based on old superstitions and outdated knowledge—a lousy guess. A monkey guessing at random once a year would have figured it out by now, but not modern medicine. All doctors—and dentists— are incompetent—it's only a question of degree.

For instance, I called up the editor of a journal named something like, "*Diabetes and Endocrinology,*" and told her that I had cured diabetes mellitus and wanted to write an article that would fit her criteria; her response was, "We're not interested." So I asked again, "You're not interested in curing diabetes mellitus?" She responded, "No, we're too busy. We have too many articles to read." This was before their first issue had been published, so I suppose she had visions of her mansion and European luxury car swirling down the drain; how horrible it would be if she had to get an honest, real job, telling the truth and possibly having to see patients, instead of perpetuating the myth of incurable diabetes mellitus! You can't pay for a Mercedes-Benz that way! Likewise, when I contact various [censored] Foundation(s) by email repeatedly over several years and tell them that I have cured Parkinson's disease and other diseases, there is no response. (I bet I would get a response if I offered to donate a million dollars, though!) Well, once I got an automated response that someone would contact me soon— they never did. The reason is because these [censored] Foundation(s) is/are not in the business of trying to cure diseases such as Parkinson's Disease and other diseases, their stated goal(s); actually it is in the business of raising money for themselves by pretending to want to cure Parkinson's Disease or other diseases, and then spending a lot of that money on themselves, and spending just enough money on ridiculous, probably pointless, research to appear to be legitimate. Oh, sure [very heavily censored] some affected people probably want(s) to cure Parkinson's Disease and/or other diseases, but he/she/they is/are out of the loop (or maybe they are just faking—many foundations have actors/actresses as spokespeople, after all); there seems to be no way to contact these spokespeople who actually have these diseases, and the people who work at these foundations apparently refuse to tell the truth perhaps because they are afraid of stopping the gravy train and having to look for a real job in a stressful economy. It's interesting . . . there are all of these Hollywood celebrities who are or were dying of allegedly "incurable" diseases, and they all have great access to the media, and yet they don't want anyone to contact them, even though the cure for their disease(s) has to come from *someone*, and when I try to contact the few who have set up foundations, they never return my call. Oh, well! What can you do?

The fact that these organizations refuse to talk to me suggests one of two things: either they know how to cure diseases, and they are suppressing the truth in order to make money, even though that is essentially murder, or they don't know how to cure these diseases, but they know that a cure will reduce advertising revenues or charitable donations, so they don't want to know, and they certainly don't want anyone else to know, either, so they suppress the truth. It's true: the media lie. I've contacted numerous media outlets—newspapers, magazines, medical journals, scientific journals, television stations, radio stations—and none of them has ever shown even the slightest bit of curiosity, even though everyone working at those places is absolutely dying of these "many diseases" that I've cured. None of them are even willing to report an opinion piece or an investigative piece such as: "A lone pariah has made a claim that he can cure (insert 50+ diseases here). Is he insane? We investigate tonight at 11:00." An editor at the [censored] newspaper told me, "That sounds too much like public service, and we don't do public service."

Perhaps the fact that your doctor is incompetent shouldn't surprise you. If you knew the history of medicine, you would know that doctors have always been incompetent, especially before the discovery of microorganisms in the 1870s or so, which is to say not only that they didn't know things (which is excusable), but also that the things that they thought they knew were wrong. Ignorance is excusable—everything has to be discovered at some point, often over and over again—but doctors doing things that kill their patients by ignoring common knowledge is less forgivable. This is the situation that we are in today. Socrates famously said:

Surely it is the most blameworthy ignorance to believe that one knows what one does not know.

Socrates said that around 450 B.C., yet people are still running around pretending to know things that are obviously, demonstrably wrong. In the 13th century, Roger Bacon noted that people believe knowledge by three different means: reasoning, experience, and authority. The fact that approximately 85% of Americans proclaim themselves to be religious tells us that, for most people, reasoning and experience take a back seat to authority. When 85% of people believe that they should follow their church's teachings without question (although not all of them do), convincing them of rational facts is a daunting challenge—these are the people who make decisions emotionally rather than logically. It is because of trust in authority that Galen's incorrect thoughts dominated medicine for 1,500 years, preventing medical progress. That's right! One thousand five hundred years! Medical thinking didn't change at all during this time frame. It has been said that if a doctor during the Galen Era happened to observe something that contradicted Galen's writings, the doctor would believe Galen rather than his own eyes. He might excuse his own findings by believing that his own experience is atypical—after all, autopsies were banned for centuries, even millennia.

For centuries, apparently, people were content to accept death as their fate, with little desire to extend their lives or to do experimentation. For centuries, religion and medicine were the same thing—medicine was just a branch of religion, because illness was believed to be caused by sin or punishment or trials from God, or by one or more gods, or occasionally by miasmas. One major difference between medicine and religion, though, is that medicine (science) offers testable hypotheses, while religion offers untestable hypotheses. Medicine can run experiments, while with religion, you either believe it or you don't—religion demands faith. Therefore, medicine is expected to evolve, while religion is expected to stay the same. (Although some people get into trouble when they join a religion and then expect it to change to suit them.) We have problems when *medicine* becomes based on *faith*, which is what we have today. When medicine is based on faith, approximately 95% of the "medicines" that are advertised on television actually inhibit peoples' immune systems and kill them (true). People have complained about the patent medicines of the 1800s, calling them names such as "snake oil" but those medicines were often good, if inadequate, medicines, in many cases, because they usually contained ethanol (alcohol), a very useful antibiotic and antiseptic; in contrast, what we have today is Big Pharma murdering people for profit, and few of their most-popular drugs are beneficial. I've heard about a drug to treat Multiple Sclerosis (MS) that costs more than $100,000 per year (I don't know which one). Obviously, it hasn't cured MS, or we would have heard about it, because Big Pharma has great public relations departments, and billions of dollars for advertising, whereas I, in contrast, have nothing for either PR or advertising. Based on the fact that the approach to curing MS involves immunosuppression, I can assure you that the drug in question is a useless, harmful drug. I've cured MS using antibiotics that cost less than two dollars a dose. Yet, because of a misplaced religious faith in modern medicine, people have refused to listen to me—apparently, they just don't believe that diseases can be cured, unless it's via genetics. People are willing to pour billions of other people's dollars into the study of genetics, despite the fact that they don't even have a solid foundation in basic health science! Build your foundation solidly before building a staircase to the stars! For most people, genetics is not the answer to their diseases, it is just a useless red herring.

Cancer is a disease that is the very definition of a genetic disease. Mysterious, it is a disease that has always been inside us and strikes suddenly, without warning. Or is it? Leukemia is a prime example, a prototypical, definitive example of cancer, so let's use it as a study subject. "Everyone knows" that leukemia is cancer, and cancer is genetic, so therefore leukemia is genetic and it has to be treated using drugs that affect the genes and chromosomes (structures that contain the genes). One method of treating a blood cancer is bone marrow transplant. The first bone marrow transplant was performed in 1956 by Dr. E. Donnall Thomas, who was later awarded a Nobel Prize. A boy came into a hospital with Acute Lymphocytic Leukemia (ALL), and he was fortunate enough to have an identical twin brother, so a transplant was performed from the healthy twin, and it was a success, and they all lived happily ever after, apparently. What should we learn from this?

Well, the most obvious fact that should jump out at you is that leukemia is not genetic! If it had been genetic, both boys would have developed leukemia, and a bone marrow transplant from one twin would not have cured the other twin—it would have been useless! With identical twins, if a disease is genetic, then either both twins have it, or both twins don't have it; one identical twin cannot have a genetic disease if the other twin doesn't. Therefore, this event should have shown doctors that most leukemias are probably not genetic. Yet, doctors continue to ignore this obvious fact, pursuing genetics as a cure for leukemia, so they can renew their grants and make tons of money, or just charge patients a lot of money. We are talking about basic science! Doctors are ignorant of the most very basic science, and yet they want to go out and explore new scientific frontiers when they don't even understand the most basic facts that every pre-medical student should know! The problem that medicine has is that the doctors who guide it are extremely prejudiced; they very often overlook the truth in order to make events and belief systems fit their prejudices, so that certain people can get more funding, so we can have a multi-billion dollar war on cancer that really has accomplished nothing, except for a few accidental successes. It is worth noting here that most successful drugs are derived from plants, the effects of which may usually be attributed to natural pesticides made by those plants.

Different special interest groups fight for billions and billions of dollars, ignoring the truth, and causing millions of people to die in the meantime. Doctors have always resisted the truth—for instance, just getting doctors to wash their hands has taken (are you ready for this?!) about 3,000 years. Currently (2017), there is a lot of excitement about a heroin addiction epidemic. I believe that this is easily treatable as an infection, but only when people finally decide to reject their prejudices. The first step in fixing modern medicine is to realize that doctors are incompetent, and that modern medical approaches are not working, and that they can never work as they are, because they are currently fundamentally wrong—the dominant paradigm (model) is false. In the pages that follow, I will explain how to cure many diseases that are currently regarded as incurable.

Chapter 3

Progressing from Seeing to Understanding

Doctors are men who prescribe medicines of which they know little, to cure diseases of which they know less, in human beings of whom they know nothing.
— Voltaire

When you think about the history of modern medicine, you probably think of someone from the 19th or 20th centuries. Perhaps you think of Dr. James Parkinson, who described Parkinson's Disease in 1817 or Alois Alzheimer who described Alzheimer's disease in 1906 or perhaps James Hodgkin who described Hodgkins disease in 1832. An important fact to keep in mind is that many of the most famous names in medicine never really did anything significant other than describing and sometimes naming a disease that had been recognized to some extent for thousands of years. The real difference is that in the 19th century, doctors broke these diseases up into microcategories, whereas before they would have just been lumped together as dementia or pleurisy or something. In fact, very few doctors have actually made significant advances in medicine. Many of the most famous doctors were not actually physicians, they were often biologists, chemists, or physicists. Examples include biologist Alexander Fleming who discovered penicillin, then neglected it only to have it modified and popularized by chemists Howard Florey and Ernst Chain, after having made it usable for treating disease. All three shared a Nobel Prize for discovering/inventing/

modifying penicillin, but for some inappropriate reason, Alexander Fleming got a disproportional share of the credit. X-rays and other types of imaging machines were invented by physicists beginning with Pierre Curie and later his wife Marie who were also awarded Nobel Prizes.

The sad truth is that physicians themselves don't seem to innovate much. Most of the innovations by physicians are probably in the area of surgery. Unfortunately, though, many of those innovations will be made obsolete by a minimally adequate knowledge of medicine that most doctors are currently lacking, and which you will have by the time when you finish this book. A great many of the doctors whom I meet seem to be dull and lacking in creativity and curiosity. When I tell them that I cured diabetes or Parkinson's Disease, not one of them has ever asked me how I did it, to the best of my recollection. That would be the very first question that I would ask if the roles were reversed! Instead, I just get dull, blank stares, or the occasional referral to a psychiatrist. It's a sad day when believing in reality is proof of insanity. Thus has it always been.

As reported by Malcolm Gladwell in his book *David and Goliath*, psychologist Dean Simonton explained the situation as being because as children, they "inherited an excess of psychological health," causing them to be "too conventional, too obedient, too unimaginative to make the big time with some revolutionary idea." I would add that this personality defect also prevents them from recognizing other people's revolutionary ideas.

Keep in mind that doctors don't have any idea about the causes of chronic diseases or how to cure them, yet to question this idiopathic 'explanation' or approach to disease that is never successful is considered to be an act of heresy. Doctors seem to be most comfortable just performing like robots. Come to work, diagnose, apply standard treatment, collect paycheck, go home. (In fact, modern doctors are the best argument for replacing doctors with computers; there would be no difference, because doctors have no intuition or insight, that I can see.) Change is stressful; nobody wants to think new thoughts! There's no billing code for that! There's no code for innovation! You can't do that! It's not allowed! You will be banned from [censored] hospital in Baltimore (true story).

The only approach that doctors use is to try to "manage" chronic diseases. Consequently, doctors seek to treat every sign and symptom separately. Diabetes mellitus (DM) is a good example of this. When insulin was invented in the 1920s, it was a life-saver for patients with diabetes, but no one thought of it as a cure. Administering insulin was just a way to keep people alive until a cure could be found. Unfortunately, though, people have gotten sidetracked onto this idea that diabetes mellitus is solely a disease of the pancreas and blood glucose. This is ridiculous! Describing DM this way is like describing an automobile as "a kind of a rectangle with a round thing." While this might be an accurate description of a car, it is nevertheless completely useless. In comparison, the current understanding of DM is only partially accurate, but it is partially useful. Using drugs to lower blood glucose (sugar) helps to keep people alive, even though it is the wrong treatment. It is one case where treating the signs and symptoms of a disease helps to keep the patient alive, even though it does nothing to treat or cure the disease. If there is any time when someone theorizes about the actual cause of DM, the usual tendency is to turn to a claim of genetics with some unknown virus, or sometimes even just a memory of a virus, which is an insane hypothesis.

The claims of genetics as a cause of disease goes back to a theory of innate degeneration—that we all just get old and die, albeit at different speeds, because we're designed that way. Since the beginning of time, disease has been thought of as something that is just inside the body, and there is nothing that can be done about it. Sometimes disease might be attributed to gods or sin or miasma, but for the most part, something like cancer was just thought to be something that was lurking inside the body and was just the unavoidable fate of every living being—all bodies would eventually degenerate in different ways. It is a case of learned helplessness. Once Watson, Crick, and Wilkens (Nobel Prize, 1962) described the shape of DNA, and once genes were discovered, this concept of innate dysfunction became embodied in the study of genes. Suddenly, every cough and sniffle could be explained by genetics and the answer to everything was just to find the right gene . . . or combinations of genes. For instance, someone discovered the gene for Cystic Fibrosis, and the future was assured for all of the Cystic Fibrosis (CF) patients . . . except then more genes were discovered, and then more, and now I think they're up to around 800 genes that "cause" CF, so I'm wondering . . . are these all competing claims? Do all of these genes work independently? Are there only certain combinations of genes that "cause" CF, and if there are 800 genes for CF and they have to appear in certain combinations to cause CF, aren't the odds of actually getting CF in this model so insanely high as to be impossible? I suspect that CF may not be genetic at all, but I haven't had the opportunity to test my ideas yet; however, the claims that are made for many diseases, especially "autoimmune" diseases are clearly illogical and insane. I guess this should not be surprising, as medicine has always been at least somewhat illogical and insane—this was inevitable, because microorganisms were not discovered and accepted until 1878, when Louis Pasteur presented his findings to the French Academy of Medicine. Still, we have so much amazing technology, these days, that we expect people to think logically, but they still don't.

Even when a gene is found, there is usually nothing that can be done to change the situation, although this is changing. Unfortunately, though, the cure is usually to completely kill the body's immune system and start over with a bone marrow transplant, sort of like turning your computer off and then back on to try to fix some problem. In contrast, it would be much, much better to be able to treat the actual cause of chronic diseases, rather than just trying to reboot and reprogram the system; rebooting the human body is a method that is doomed to failure, or at least inadequate success.

In reviewing the history of medicine, it is really not all that clear how successful it has been through the ages. At times, such as prior to 1850 or so, many surgeons had failure rates of 80%, which is to say that 80% of the patients died. Today, such a doctor would quickly be put in prison, but back then, no one had any idea what caused diseases, and death was just accepted as God's will. Microorganisms had not been discovered, and the few people who theorized about "animalcules" were ridiculed and shamed. During plagues that recurred throughout Europe for thousands of years, more than half the residents of some cities died over a period of weeks, and most doctors fled to the countryside with everyone else. Medicines consisting of herbs and minerals and various other substances were used to treat various diseases, such as mercury or arsenic to treat syphilis, but it is unclear what the difference was between herbalists and doctors; also, physicians were considered separate from surgeons. As muddled as the history of medicine is, one thing is clear: it didn't work very well, if at all.

The great improvement in the success of medical treatment began when Louis Pasteur proved the existence of microorganisms—bacteria—starting around 1865 or so, and continuing until at least 1878, when Pasteur made his presentation to the French Academy of Medicine. Still, decades more were required to bring doctors around to the new science, and there was still little or nothing that could be done to treat bacterial infections until the late 1930s and, especially, 1941, when penicillin became available. Thus began an era of bacterial treatment. This led to viruses, fungi, and various other microorganisms, and yet the many diseases that were identified and named before the era of microorganisms continued to be thought of as diseases of innate dysfunction (e.g. Parkinson's), for reasons that are obscure and undoubtedly illogical. It may be impossible to say exactly why the thinking about such diseases as cancer, which has been written about for 3,000 years, never changed. It may be that some diseases were just believed to be associated with old age, and the sciences of genetics (and viruses) came along at just the right time in order to appear to be the logical subsequent course to pursue in order to understand old age. Unfortunately, doctors jumped too far too soon, and it is now clear that many diseases that had been thought to be due to old age are actually due to infection by microorganisms that are different from the ones that people usually think about. In a great many cases, genetics is a red herring, a train running off the rails through the meadow while the engineer keeps piling on coal.

As one example of the misapplication of genetics, consider that first bone marrow transplant, which was performed in 1956, mentioned earlier. Despite the fact that only one twin developed leukemia, doctors still erroneously consider ALL to be a genetic disease, which is insane! This case is definitive proof that ALL is not a genetic disease, because if it were a genetic disease then both twins would have had it, and a bone marrow transplant would have been useless. Yet doctors have spent 60 years so far ignoring this inconvenient proof because it doesn't fit their models of disease; doctors are very selective about what evidence they will believe— only the "facts" that fit their prejudices.

The problem with all of this is that there is a class of infectious organisms that have been recognized for millennia, with one specific organism in particular that was discovered in 1876 by Louis Normand seeming to fit the paradigm with perfect consistency, but doctors fail to consider it for reasons that are essentially racist or at least borderline racist. And there is one other hidden factor that doctors have also failed to understand, one that won a Nobel Prize in 1960 . . .

Chapter 4

The Cause of Many Chronic Diseases

Talent hits a target that no one else can hit. Genius hits a target that no one else can see. —Arthur Schoepenhauer

So if doctors are almost completely wrong, and if the innate—or genetic—explanation of disease is often wrong, what is the cause of so many "incurable" diseases? There's a guideline, an ancient dictum, that doctors are taught in medical school called Occam's Razor, which states:

The most likely explanation is the simplest explanation.

If you apply this concept to the many chronic diseases that we're facing today, the most logical approach to try is that of infection—it is very simple. These many chronic diseases that we're facing today are the result of infection. Infection fits the paradigm perfectly! It's an infection!

I used to live in a redwood forest. The house that my mother and I lived in was an old summer cottage that was built in Mill Valley, California, before the Golden Gate Bridge was built. It was built on a very steep slope that ranged from about 25 degrees to vertical. The house was essentially square, and one of the uphill corners, the northern corner, was built on a redwood stump that was left in the ground. The house was around 70 or 80 years old, and the stump received little or no direct sunlight. One day, some of the shingles that covered up the stump fell off or got knocked off, and after a few weeks, a new sprout appeared on the stump. This redwood tree, that had been cut down, covered up and hidden from the sun for 70+ years, was still alive! Redwood trees are one of the longest-lived trees—and organisms— on earth, living more than 2,000 years, but this just seemed astounding! So then, that raises the question: What determines long life? The answer is that long life results from an immunity or resistance to disease/infection. Redwood trees seem to be immune to every disease that kills other trees. There are only a few things that seem to kill them: people, drought, high winds, and fire—and even death by fire isn't all that certain, as a surprising number of redwoods have been hollowed out by fire so that they look like a giant drinking straw, open to the sky, and yet they continue to live! The droughts that may kill some redwoods tend to be caused by the effects of human activity on global climate change, so mostly what kills redwoods is people. People can be thought of as an infectious organism that kills trees, in a loose sense. Extrapolating from redwood trees, it seems then that there are only three causes of death in general: trauma, starvation, and infection. For people and pets, the main killer is infection.

This concept of long life in trees being a result of immunity to disease also applies, albeit somewhat less well, to other organisms, including mammals, and therefore humans. I recall reading somewhere that in the absence of negative factors, the human heart could beat for 600 years. Now, I don't know if people can live 600 years, but I don't know that they can't, either (although I do know that they don't). What I have noticed is that all people suffer from disease, and old people obviously (to me) die of infection, not old age. Everything that you think of as old age is actually due to a parasitic infection—loose, sagging, wrinkled skin, arthritis, failing eyesight, urinary troubles, heart disease, lung disease, kidney failure, and many other diseases. The paradigm that fits best is parasitic roundworm. There is absolutely no reason to think that any mammal has ever lived without parasitic worms, as they are ubiquitous in the environments where people have lived basically forever.

The first reaction that most doctors—and perhaps most people—have to the claim that literally everyone has worms is that that is crazy, impossible, ridiculous. This reaction, I contend, is proof of the extreme prejudice that prevents people from even considering parasitic worms as a cause of disease and aging. There are basically three possible responses to this or any claim: The common response is "that's crazy/impossible/ridiculous!" This response is irrational, because the claim is demonstrably plausible, as is evidenced by the major medical reference books that describe these infections. People who make this response are ignorant at the very least, and probably stupid, too. The second response is, "That's not in the literature." This is a way of saying that the doctor isn't willing to think; he/she wants other people to do his or her thinking for them. This is really a very pathetic, stupid, and tragic way to live. This person is just a robot who should be replaced with a computer, because a computer can be reprogrammed, whereas the doctors at [censored] Hospital in Baltimore, as one example, apparently cannot be reprogrammed to think. The third response is "What's your evidence?" This, or some version of it, is the only rational response to any claim, no matter how ridiculous or seemingly insane. Ninety-nine percent of doctors to whom I've spoken have failed to respond rationally, instead defending their indefensible prejudices that often have their origins in ancient, primitive, pre-scientific, semi-religious medical beliefs. This suggests that doctors are no more logical than any other person; physicians tend to be just as irrational as the average person, which is to say, "very irrational." Would you prefer that your doctor be rational or irrational? Most people want a doctor who thinks logically. Most people don't receive that, though.

Most people think that doctors are very smart, but this isn't really true. Education tends to be about memorizing facts, and the people who become doctors tend to have superior memories. Although doctors tend to be very good at memorizing information, this is not the same thing as being smart—it is only one part of being smart. Because education is largely about memorizing facts, people who can memorize a lot appear to be smarter than other people (and they are, in a way—but only to a limited extent). However, memory is only a part of intelligence; the most important aspect of intelligence is being able to think and solve new problems. Medical students, however, as a general rule soon become overwhelmed in medical school, so they quickly change from a type of learning that seeks to understand to the type of education that uses rote memorization (memorizing without understanding) to get through medical school. They start to learn like robots, and people who learn like robots don't like to be reprogrammed. Because they don't understand what they're doing, they are too uncomfortable to change their programming; if they can be forced to understand, they could resolve this cognitive dissonance. People who understand things are more willing to change their minds when events become unsatisfactory, whereas soldiers who just do what they're told are very resistant to changing their behavior—or, at least, this is a general theory. A great many things about modern medicine don't make sense, but doctors persist in the old ways out of habit and unwillingness to change. A great example of this is the percussion (tapping) of the chest with the fingers.

I was doing a clinical rotation with a doctor, and, like most doctors, he would do percussion of the chest as part of his exam before listening with his stethoscope. I asked him if he had ever discovered anything in a patient as a result of percussion, and his response was, "Never." This man had approximately 20,000 patients, so performing a useless procedure on each one must constitute a considerable waste of his time, so I asked him why he bothered to do it if it had never been useful in 20 or 30 years of practicing medicine. His response was, "Because I'm supposed to." The problem here is that percussion of the chest was invented in 1761, while the stethoscope was invented in 1816 and, while the stethoscope has evolved considerably since its invention, percussion has stayed the same. As far as I can tell, the stethoscope should have made percussion completely obsolete, and yet doctors still use it, despite its complete uselessness—it is a technique that has been obsolete for 200 years, yet doctors continue to use it because they think they're "supposed to." Medicine is a profession that abhors change.

Most doctors are completely ignorant of parasitic worms, largely because their teachers spent about an hour on the subject in medical school, and because there is usually only one worm-related question on the licensing exam (out of about 600 questions)—but this usually doesn't stop them from proclaiming themselves to be experts. The average American doctor has never diagnosed a parasitic worm infection and refuses to consider it. When I told one doctor whom I was seeing as a patient (i.e. I was the patient) that I've been treating myself for worms, her first words were, "I strongly urge you to see a psychiatrist." Now, that's prejudice! (I have a Master's degree in Psychology, among my seven college degrees.) In modern American medicine, it's o.k. to diagnose someone with "delusion of parasitosis," but not actual parasitosis (infection with parasites), because Americans (and Europeans) are considered to be just too good to get worms, apparently, despite the fact that parasitic worms are endemic worldwide. Generally, the unspoken belief is that parasitic worm infection happens only to dirty little brown people in third world nations who live on trash dumps or work barefooted in rice paddies. Five minutes on the internet would prove that worms are found worldwide, but doctors can't be bothered with the truth. They can't handle the truth.

It's important to recognize that there are literally thousands of different kinds of parasitic worms, and official estimates of the number of people infected with different worms add up to around 7 billion. The worms that most people will immediately think of are the 12-foot-long intestinal worms that they may have seen in some newspaper story or in some record book, but these are rare exceptions and are not the worms that affect most people. The worms that affect most people are usually less than 2-3 millimeters in length, and they may be primarily microscopic, possibly being so small that they can live inside a single human cell. Many worm eggs are much smaller than a single cell, so this is not unreasonable. A worm such as *Strongyloides stercoralis* can infect any organ in the body, and reproduces so slowly that it appears to be "just old age." Parasitic worms are the cause of arthritis and many cases of pain of unknown origin such as back pain and plantar fasciitis (foot pain), and varicose veins in the legs. These worms tend to gravitate downward, so they end up in the feet, the legs, the pelvis, and the wrists and hands, but they can also get into the brain and spinal column. Parasitic worms are the leading cause of visual problems and dental decay, but these facts are not recognized by the medical or dental professions.

In an informal survey, I've found that most people who wear glasses have "floaters" (aka scotomas) in their eyes, which are spots that move around, and I can see that the floaters in my own eyes are worms—very tiny worms, which Dr. [censored], a Baltimore eye doctor, just laughed off, completely dismissing my objective observations despite the fact that I probably have more education than he does. The best way to see these worms is to go outside to a meadow or field on a clear, sunny day. When lying down and looking up at a clear blue sky, you may be able to focus on your floaters, if you have them, and you almost certainly do if you're older than 40. They move away from your central vision when you try to focus on them, and they often cannot be seen clearly. They often appear as masses or tangles of worms that are indistinguishable, but every so often, you might be able to see one or two clearly, if you are patient.

Eye doctors have a variety of truly insane explanations that they use to explain floaters, such as, "cracks in the vitreous humor," which is the jelly-like substance in the eye. I've never seen Jell-O™ crack, and if it did, it certainly wouldn't move around. This is just another example of doctors continuing to use primitive, ignorant hypotheses long after they should have been discarded. The eye doctor in my insurance plan was a very obese, bald man wearing Coke-bottle glasses who refused to take me seriously. He had no interest in preventing my vision loss; he just wanted to prescribe me glasses every time that my vision gets worse. (I still don't wear glasses.) I guess that prescribing glasses must be a very lucrative business. Here's a clue: If your eye doctor wears Coke-bottle (i.e. thick) glasses, he almost certainly can't help you to preserve your vision, either. If he can't save his own vision, why should he be able to help you, even if he wanted to, which this doctor obviously didn't? And if he's fat and bald, then he really doesn't have much useful health information, because these conditions can be prevented, or at least mitigated. I've never seen a doctor who looks as healthy as I do, even though I'm not nearly as healthy as I'd like to be. I have a policy: I don't take health advice from any doctor or person who is obviously sicker than I am—and all other doctors are obviously sicker than I am! (So far!) This quack eye doctor told me that, "a lot of people report seeing long slender objects, threadlike objects, or snake-or-worm-shaped lines in their vision, but those aren't worms." This man is so delusional that he has "many" patients who tell him that they see worms in their eyes, and he just automatically rejects them one hundred percent of the time, with only the most-superficial of exams! He is a very stupid, arrogant man, and there's nothing that the average patient can do about it except to go elsewhere, which insurance doesn't usually allow, so patients are just fastened by an inclined plane wound helically around an axis (screwed)!

Most people seem to have an initial reaction that they couldn't possibly have worms. No matter what a person's background and education are, they think that they know all about medicine, and they certainly know more about medicine than I do. It doesn't matter if their college degree was in music, or fine arts, or business, or law, everybody thinks that they know more about medicine than I do! This is another example of common-but-insane thinking. People often have irrevocable opinions that are based on a foundation of total ignorance. It is inexplicable.

I have cousins who live in North Carolina, and they don't think that they could have worms, even though the John D. Rockefeller foundation spent a great deal of time and money in the 1920s or thereabouts to fight hookworm in residents of the southern states with an emphasis in North Carolina. Of course, this happened long before my cousins were born, so as the old knowledge dies out, the pattern repeats itself, although not as strongly as the first time.

Everyone seems to think that they have never been exposed to parasitic worms—they've never played in the grass or hiked in the woods, never eaten a vegetable, never been swimming in a lake or creek. It doesn't matter—this is the genius of these worms. You would still have a worm infection if you had been born and raised in a sterile bubble. The reason is this: these worms penetrate every organ in the body, and they are passed from generation to generation in the womb. This means that they penetrate the uterus. They travel in blood and lymph, so it's impossible that they are not exposed to the uterus and the placenta in the nine months required to grow a human being in the womb. Once exposed to the placenta, they travel down the umbilical vein into the liver of the fetus, where they start a new cycle in a new host organism. However, perhaps the most intriguing twist is that because the worms are introduced into the fetus before the immune system develops, the worms are never recognized by the host's body as being foreign, so the immune system never develops an adequate, or perhaps any, response. The people who first noticed this phenomenon, MacFarlane Burnet and Peter Medawar, were awarded the Nobel Prize for Physiology or Medicine in 1960. However, they were not concerned with diseases/infections, they were doing research for transplantations—they probably never even thought about infections, much less worms; they were hoping to grow new body parts which could then be transplanted into someone. I don't think that they really found what they were hoping for, but their research nevertheless explains how parasitic worms are passed from generation to generation, appearing to be the result of genetic inheritance, without the true infectious connection ever having been noticed before.

When you combine this work with the work of Johannes Fibiger, who won the 1926 Nobel Prize for Physiology or Medicine, by showing that parasitic roundworms can cause cancer (see chapter 15), one wonders why no one else figured this out paradigm earlier. Worms can cause cancer and so-called autoimmune diseases (which are just diseases with no obvious cause—not the body attacking itself); worms love the lymphatic system, which explains lymphomas; they travel throughout the body and thus can cause disease of any organ system—even the spinal cord (think: Lou Gehrig's disease [ALS], which I've cured)—all without being recognized by the immune system. And because the worms in question, such as *S. stercoralis*, multiply so slowly, it all appears to be "just old age."

Many, many diseases are caused by parasitic worms, including Celiac Disease/Gluten Allergy, Irritable Bowel System, and many others, which I've cured. These diseases are falsely labeled "autoimmune" diseases, and they are "treated" with immunosuppressive drugs such as Humira ™. Again, consider the warning that appears in every Humira™ television commercial:

Humira can lower your ability to fight infections, including Tuberculosis. Serious, sometimes fatal infections and cancers including lymphomas, have occurred, as have blood, liver, and nervous system problems, serious allergic reactions, and new or worsening heart failure.

"Oh, yeah, man, I wanna get me some of that!" Is that what you're thinking right now? Probably not. The drug company might as well just say, "Autoimmune disease is a scam, and this drug will allow your parasitic worm infection to explode and kill you." However, Humira™ is not alone—as I said earlier, about 95% of the drugs that are advertised on TV are just as bad. Some doctor prescribed a drug called Breo™ for a friend of mine, and the next time that I saw him, he was an invalid in the hospital. I now expect him to die soon. Coincidence? Not in my book! The warning said that Breo™ increases the risk of hospitalization and death. Why would anyone prescribe or use such a drug? Stupidity is the only reason that I can think of. I've cured several so-called "autoimmune" diseases that were actually misdiagnosed worm infections, leading me to believe that autoimmune diseases probably don't exist at all—it's just an antiquated concept from a pre-scientific era. I've cured Parkinson's disease, multiple sclerosis, Lou Gehrig's disease, Irritable Bowel Syndrome, chronic pain, and many other diseases, including some cases of Diabetes Mellitus, and I see nothing to indicate that autoimmune diseases are real, although I can't say definitively yet that autoimmune diseases never happen, just that all of the ones that I've looked at are not real or were misdiagnosed. There are supposed to be more than 80 "autoimmune" diseases, and I believe that they are all misdiagnosed parasitic infections.

Let's think for a moment about infection. Let's make an analogy between infection and World War II. In this analogy, England and the Allies—even Europe—are the host body, while the Nazis and the other Axis powers/ soldiers are the infection. War, or battles, is analogous to inflammation. There are two ways to reduce inflammation (battles or conflict): Either let the Allies win, or let the Nazis win. If you let the Allies (the immune system) win, inflammation is reduced and life goes on normally. If you let the Nazis win, inflammation is also reduced, at least at first, or at least on the surface, but soon the Jews are quietly sent to the gas chambers, then various other groups (e.g. trade unionists) are killed off slowly, one-by-one, and the organism (Europe) dies very slowly, but only after a long period of "peace," or what appears to be peace—a long, slow period of slow deterioration (low inflammation) and decline known as old age, which, obviously varies considerably in terms of length depending on one's health status and financial situation.

So, when you take immunosuppressive drugs to treat your pain or disease or whatever, it "works" (relieves pain) only for a little while before it kills you by allowing your worms (Nazis) to thrive. This is why shoulder or knee pain is sometimes treated by one or two shots of hydrocortisone, but not three shots, because three shots always make the disease noticeably worse. Doctors still don't have a clue why. The reason should be obvious, but doctors don't read books—such as medical reference books—anymore, and the internet isn't really such a good source for medical advice, because it is all profit-driven, and powerful people censor people like me who try to distribute the truth, because they make more money when people do more internet searches, and sick people do more internet searches for health matters than healthy people do, so search engines including [censored] want to keep you sick (When I wanted to market a website on Google, I was told that they prohibit the use of the word "cure," presumably because they don't want people to learn how to cure their diseases—it sounds like genocide to me), just as medical journals don't want diseases cured, because then there is less or no need for medical journals, and no need to advertise in medical journals, reducing or eliminating revenues. Drug companies have media outlets in their pocket, because they are some of the biggest advertisers, so don't expect the truth to come out too soon, and don't be surprised if the media fight the truth to the bitter end.

As another example of the media resistance to the truth, I personally put some of my unpublished articles into the hands of [censored], the anchorman who was advertised as the head of [censored] News, and told him that this is the biggest story that he could report in his lifetime, and, as you can see, he has chosen to continue to become very wealthy and let people suffer and die rather than to report the truth and decrease his income from drug advertising revenues. This is despite the fact that [censored] News makes a big pretense of being investigative journalists, when, in fact, they pretty much just report the same stories that they receive from their news service, probably the Associated Press, the same one that supplies all of the other news networks. If you watch two or three network news shows side-by-side, you'll see the same stories in almost exactly the same sequences, night-after-night, except for the last—human interest—story. Rarely do the main stories vary from one network to another.

As another example, I went on a website called mydailystrength.com, and said that I had cured Parkinson's Disease. They immediately banned me from the website for life. So this website is another example of people who profit by censorship, by keeping people sick, by preventing the truth from coming out—but if you want to promote a really expensive patented drug that will bankrupt people without curing them, they will welcome you with open arms! That reminds me of something that an editor at the *Journal of the American Medical Association* (JAMA) told me; he said, with surprising grammatical correctness: "We are interested only in articles about patented drugs." In other words, they're not concerned with good medical treatment, only expensive medical treatments. Huh. It's more [censored] for fun and profit. Well, at least he was honest enough to essentially come right out and say that he is a [oh, very, very censored]. Most [still censored] are not as honest as this JAMA editor. Modern medicine is a scam, but many doctors are also victims as well as perpetrators.

Chapter 5

Why Medicine is So Expensive

There are three kinds of lies: Lies, damned lies, and statistics. —Mark Twain

Medical care is more expensive today than it has ever been before. Do you wonder why? It is because all doctors are incompetent, just as they have always been incompetent throughout history—they are giving the opposite of the right treatment and they are ignoring the major infection on earth, parasitic worms. The difference now is that never before have doctors had such an array of drugs that suppress the immune system in such new and clever ways—in fact, this is the first time in history that doctors have ever *wanted* to suppress the immune system. These drugs make people sicker, because everyone is *infected* by parasitic worms that have been passed from mother to fetus since the beginning of time; this applies not just to humans, but to all mammals, and most likely all other species. (I'll bet that the breeding problems of giant pandas are caused by parasitic worms.)

When you have an infection, you need to make your immune system stronger, not weaker, in order to fight off the infection. Today, doctors are doing the opposite, deliberately trying to weaken your body by weakening the immune system with these new immunosuppressive drugs. Pretty stupid, right?! So why do they do it? The answer is that sometime before or only shortly after microorganisms were discovered, someone came up with the (false) idea that because disease is innate to the body, the body therefore should be suppressed.

In the 1930s, the immunosuppressive drug cortisone/cortisol/hydrocortisone was invented/isolated, which allowed this theory to be tested and it seemed to help to treat arthritis, seeming to support the erroneous innate disease theory. Keep in mind that doctors see their patients for only a few minutes, and may see them only once or twice, ever. Therefore, the doctors thought only to the first stage of analysis, and then they stopped. They weren't especially interested in the cause of the disease, because they had unshakeable prejudices, unalterable religious-belief-like faith in a ridiculous medical model—they knew "for certain" that people's bodies just attack themselves at some point. They don't need no stinkin' evidence! If the patient didn't return for another appointment, then "the treatment must have worked," and the long-term consequences were never considered. Just try to remember or imagine how primitive medicine must have been in the 1930s, and think about how unlikely it is that medicine could do anything at all properly. Remember that World War II analogy that we discussed earlier? Well, cortisone/cortisol/hydrocortisone is analogous to the Nazis—it brings "peace" for a while, but at the cost of lingering, chronic disease and premature death by slow, festering infections (the gas chambers).

Part of the problem here is that only one hypothesis was tested, and not its opposite. They wanted to test whether suppressing the immune system would help treat arthritis, when what they should have been testing is: Which produces a better long-term result: immunosuppression or antibiotic therapy? Unfortunately, there were no antibiotics in the early 1930s, so all they could do was to test immunosuppression, which seemed to be effective, at least at first, thus sending medical science off in the wrong direction. If you want to test whether suppressing the immune system improves a disease, then all you have to do is to use the immunosuppressive drug and see what happens. In the case of cortisone, short-term relief occurred, so the experiment was deemed successful. Unfortunately, the experiment didn't last long enough to discover the hidden negative effects, which show up only with prolonged treatment—such as osteoporosis and death. In any event, the opposite contention—was antibiotic therapy better?— was impossible to test at first, and when it finally was possible to test antibiotics, they were still much, much more difficult to test accurately because they had to be targeted so precisely: if you want to see if the disease is due to an infection, and if treating the infection might work better than wiping out the immune system, then you have to first identify the type of infectious organism from among a dozen or more general types, then you have to narrow your choice down to one specific type out of literally thousands or tens of thousands, and then you have to find a relatively specific antibiotic that will work against that particular organism, at least a little bit (there are different degrees of effectiveness). Because the first antibiotic (Penicillin) didn't go on sale until 1941 or later (although Prontosil, a bacteriostatic drug, may have been available in 1939), there were no antibiotics in the early 1930s, so in the early 1930s, there was only one medical test to be done: does immunosuppression help? Answer: it seems to help, at least at first. Consequently, people got in the habit of using immunosuppressive drugs, and they stopped asking what the cause of these diseases was, because "everyone knows" that the cause is innate dysfunction, as "proved" by the results that were obtained after giving immunosuppressive drugs in the inadequate drug trials. When penicillin finally came on the market, after a decade of using immunosuppressive drugs, people weren't going to abandon their old, familiar habits—they love their old drugs! And everybody just assumed that those arthritic, gout-ridden people who shrank, shriveled up, developed hunchbacks and scolioses, and turned ghastly white and gray would have done that even if they hadn't been on immunosuppressive drugs—which might be true, but without the drugs, they might have lived years longer before those things happened. It's another untested hypothesis.

Often, when treating a disease with immunosuppressive drugs, one problem or set of problems appears to resolve, and then later "some other disease" pops up in a different body part or area. Doctors generally assume that there are many diseases that are different because they affect different organ systems, when in fact many diseases have only one cause—parasitic worms that hide their identity by traveling to different organs or parts of the body. (And those "spider veins" that are so common are almost-certainly not veins—they're almost-certainly larva currens or larva migrans, a.k.a. parasitic worms or their tracks.) Doctors have no problem thinking that a bacterial infection can travel through the body and appear in different places, but because diseases such as arthritis are considered to be just an innate dysfunction, it is supposed to stay in one place, one part of the body—which is ridiculous! For instance, if you get a joint pain in one extremity, and you get a rash somewhere else, doctors will often treat them as separate diseases, even though they might have the same microorganism as a cause. One disease might get an antibiotic, while another body part gets an immunosuppressant, or both diseases might get immunosuppressants. Then, after you have been taking immunosuppressant drugs for 20, 30, or 40 years, you contract a bacterial infection that you can't fight off. Why can't you fight it off? It's because doctors spent decades, or maybe even just a few weeks, destroying your immune system! When you combine doctors' efforts to destroy your immune system with the natural harmful effects of parasitic worms on your immune system, you can see that everyone who doesn't die prematurely from trauma ends up having a suppressed immune system, and this causes them to turn into shriveled, little, old, gray people. A big part of the reason for this is that worms use up minerals, and minerals are critical parts of the immune system. Two especially important minerals that get depleted as a result of chronic worm infection are copper and selenium, which are essential components of copper-zinc superoxide dismutase and glutathione peroxidase, respectively—essential components of the immune system and detoxification system. Copper deficiency in particular leads to gray hair and wrinkled skin.

A common fatal infection is caused by bacteria named MRSA (methicillin-resistant Staphylococcus aureus). Healthy people can fight it off, but unhealthy people can't, so they die, sometimes. What distinguishes the two? One factor is clearly whether a patient has been given immunosuppressive drugs for years or decades. The other main factor is usually considered to be age. The problem with this is that virtually everything that you think of as old age is actually caused by parasitic worms, because I've been able to reverse age-related decline by treating for worms. (No one has ever lived without having parasitic worms, so no one has ever seen parasite-free old age—we don't know what it would look like. Yes, I know that this blows your mind, but there is no other logical conclusion.) Parasitic worms are actually the main cause of immunosuppression, not old age. I have demonstrated this by reversing immunosuppression, gray hair, facial wrinkles, fat accumulation, muscle loss (sarcopenia), emaciation (cachexia), osteoporosis, spinal degeneration, chronic pain (especially of the back and pelvic/sacral/lumbar/gastrointestinal areas), gastrointestinal problems, dental deterioration, and visual degeneration, etc. that were alleged to be purely the result of old age—simply by adequate treatment of parasitic worms in combination with some nutritional therapy. These diseases are only the tip of the iceberg when it comes to diseases caused by worms. Parasitic worms can infect any organ, even bones, so the possibilities are almost endless, including leukemias and other cancers. Any good medical reference book will verify this. Speaking personally, I quite possibly look younger now than anyone else who is my age, because I have virtually no facial wrinkles except for 1½ lines on my forehead. (For instance, I think that I look about 20 years younger than Dr. Oz, even though he is six months younger than I am.) Also, some crippling plantar fasciitis (foot pain) that I had for 40 years has recently cleared up. This pain was severe enough to prevent me from doing a great many activities that most people take for granted, such as going to shopping malls, or any place where a long walk would be involved.

So then, remember we were wondering why medical care is so expensive? The main reason is because doctors are doing the opposite of what is beneficial for you. They are suppressing your immune system when they should be strengthening it, instead. Sure, insurance companies jack up the price, too, but they are also paying incompetent doctors who are still largely practicing early 20th century medicine. They are killing you with immunosuppressant drugs. You are paying your insurance company, your hospital, and your doctor to kill you and rob you blind, but you're so sick that you can't fight back (indeed, they often keep patients incapacitated with anti-psychotic drugs), and you don't have the knowledge to save your life on your own anyway, and you're desperate for help, and there is no one else to turn to, especially when the hospital denies access to anyone who has an intelligent opinion (e.g. [censored] of Baltimore), for fear of a lawsuit. Then they have to attempt to compensate for the damage that they have done to your health, but they don't even have any awareness of what they've done, so if you get better, it's really just almost-pure luck, combined with intravenous fluids and a very small amount of white bread nutrition and sodas with high-fructose corn syrup. The more they treat/kill you, the more "treatment" you need, the more it costs, the more they charge you, the more they laugh all the way to the bank.

Doctors refuse to acknowledge the existence of parasitic worms even though they are described in every medical reference book (doctors don't read books anymore, I guess—all of their training comes from the internet.

Does that reassure you? It horrifies me!). These worms have been recognized for 140+ years (since 1876), and literally every person has them. The thing that is new here is that I am the first person to apply the concepts developed by Nobel Prize winners McFarlane Burnet and Peter Medawar, who showed that the fetus never recognizes foreign antigens (tissues), even after birth and throughout life. Therefore, your body is literally unable to respond to your worm infection, because it thinks that the worms are just part of your body. (Osteoclasts, perhaps?) It's really a brilliant infection; you have to admire its cleverness, or good luck.

Doctors adamantly refuse, over and over again, to admit that parasitic worms exist, or exist in the U.S., or something—I'm not sure what their crazy beliefs are. I can't tell you how many hours I've wasted arguing with clueless, stupid doctors who always look much, much sicker than I do. (I have a policy: I never take medical treatment advice from any doctor who looks sicker than I am. Consequently, it has been a long, long time since I've taken any doctor's treatment advice. I will take advice about running tests, though.) Doctors adamantly refuse to use the right treatments, and they actively prevent others from using the correct treatments, so not only are doctors useless, they are actually worse than useless. Doctors are killing us. (You more than me.) And for killing you, they are willing to take literally all of your money, even your home, if you don't pay their insanely outrageous bills.

The correct treatment for these diseases costs less than five dollars per day. How much does advanced treatment in a hospital typically cost? Thousands of dollars per day. A drug that allegedly helps-but-doesn't-cure multiple sclerosis (MS) costs more than $100,000 per year—just for one drug! I have *cured* MS for slightly more than forty cents per day. When will people recognize that modern medicine is a scam?!!! Recently, it was reported on TV that the price of one asthma drug (albuterol) increased from $11 to more than $2,000! That's just evil! In some cases, people are actively suppressing the truth about diseases, and in other cases doctors are just too stupid and arrogant to change their minds. Doctors die of worms, too, so it's not like they actually seem to know the truth—they just cannot stand the thought of changing their minds! They cannot tolerate being confronted with new ideas, just as they would never consider changing their religions. Perhaps their egos are just too fragile. They have the intelligence of small children, but with a lot less curiosity. Plus, they're making a ton of money;

if their pay were performance-based, they might be more interested in the truth, but for now, the longer they can keep you sick, the more money they make. The sicker they make you, without killing you, the more of your money they can take. They have a financial incentive to keep you sick. And of course, when you spend more time in the hospital, a lot of other people get to be paid, too: workers, nurses, therapists, security guards (who help to keep the truth out), administrators, executives, etc. It all costs a lot of money—money that you would not have had to spend if you and your worm infection had been treated appropriately. After your hospital visit, they will often send people to rehabilitation hospitals that they probably have financial interests in, too.

As I mentioned earlier, *Diabetes and Endocrinology* wasn't interested in curing diabetes. Likewise, an editor at the *Journal of the American Medical Association* (JAMA) told me, "We are interested only in articles about patented drugs." I was impressed by his grammatically correct use of the word "only", but not by the fact that he, in my opinion, preys on the weak and the sick by suppressing the truth that would end their misery. Similarly, *The New England Journal of Medicine* also rejected my articles without even looking at them. Apparently, they consider any scientific evidence involving less than 10-to-20-or so subjects to be "anecdotal," and therefore contemptible and useless. Also, I have repeatedly contacted one/several [censored] foundation(s), which exist(s) allegedly to cure Parkinson's disease (as one example), and told them that I have cured Parkinson's disease and other diseases. They have never contacted me, even to discredit me; they don't even want to acknowledge my existence. Apparently, they much prefer to collect lots of money from people and live the good life. Curing Parkinson's disease [and other diseases] would put an end to what might be—or could theoretically be—construed by some people to be a fabulously lucrative scam, so it appears that [censored] they might be working hard to suppress the truth; it appears that they don't care how many [censored] people die—the organization will long outlive them. We've survived without Robin Williams, who died of Parkinson's disease, so I guess we'll survive without [censored], those other people, too. Sorry, [censored-name-person or persons], I'm trying to help you, but powerful people are silencing me [just notice how censored these chapters are—what ever happened to my First Amendment right to write the truth?], and I'm not rich enough to overcome their malicious interference. They don't want an actual clinical trial, because that would end their party. There is no reason to believe that anyone will do a clinical trial about whether worms or flukes or other parasites cause these diseases, because there is no patented drug or surgical robot involved, thus there is no profit motive. The only motive is altruism, and that ain't gonna do it for the tycoons who rule the world—it's just insufficient motivation—they care about money more than life itself.

As you can see, modern medicine isn't designed to help sick people; it's designed to make money for certain privileged people. Curing diseases would make the field of medicine much less profitable, unless they can convince the public that the only way to cure a disease is with a fabulously overpriced, patented drug. If you have a fatal disease that is cheap and easy to cure, don't expect a doctor to tell you, because he/she can't make any money off of you in that case.

People need to take control of their own health, because no one else is going to care enough to help, at least not in the foreseeable future. It's not necessary to understand everything that doctors are supposed to know— even doctors don't know those things. Here's the proof: If you go through a medical reference book, such as the *Merck Manual* or *Current Medical Diagnosis and Treatment*, there are almost-certainly more than 100 diseases that are considered to be idiopathic (of unknown cause). This includes most of the big, important diseases such as cancer; diabetes; heart disease; lung disease; gastrointestinal problems of every size, shape, and color; strokes; atherosclerosis; etc. Doctors sometimes pretend that they understand these diseases, but that is impossible because they don't actually know what causes them! Doctors can't even understand this simple concept! They don't know that they don't know! This is the definition of being delusional! "Delusional" is when your understanding does not conform to reality. Delusional is when your treatment methods consistently end with the patient's death, and yet you're unwilling to try any other method to save your patients. Delusional is when doctors routinely appear to be almost as sick as (or sometimes sicker than) their patients. All of these things are true. If I sound exasperated, it is because doctors' stupidity and intellectual rigidity are so . . . exasperating!

In the next section of this book, I present some more-formally written scientific articles, most of which were rejected without even being submitted, much less read. So much for open-minded inquiry in the scientific literature.

SECTION 2

Scientific Articles

Chance favors the prepared mind. — Louis Pasteur

The art of medicine is to cure sometimes, to relieve often, to comfort always. — Ambroise Paré

I became convinced that the physician who earnestly studies, with his own eyes— and not through the medium of books—the natural phenomena of the different diseases, must necessarily excel in the art of discovering what, in any given case, are the true indications as to the remedial measures that should be employed. — Thomas Sydenham

It is thus with regard to the disease called Sacred [epilepsy]: it appears to me to be nowise more divine nor more sacred than other diseases, but has a natural cause from which it originates like other affections. Men regard its nature and cause as divine from ignorance and wonder, because it is not at all like to other diseases. And this notion of divinity is kept up by their inability to comprehend it, and the simplicity of the mode by which it is cured, for men are freed from it by purifications and incantations. But if it is reckoned divine because it is wonderful, instead of one there are many diseases which would be sacred. — The Hippocratic Corpus

Chapter 6

Erreurs Populaires – Delusions of Modern Medicine

I know better perhaps than another man, from my knowledge of anatomy, how to discover disease, but when I have done so, I don't know better how to cure it.
— Matthew Baillie, from *The Morbid Anatomy of Some of the Most Important Parts of the Human Body* (1793)

Summary: Ancient prejudices have prevented medical progress, even today. By correcting these prejudices and introducing logical thinking, we can eliminate false attributions of disease; this then leads inexorably to curing diseases that are currently considered to be incurable. History is a useful tool for seeing wrong turns in medical thinking, and chemistry combined with infectious disease explains more than 50 "incurable" diseases perfectly. Autoimmune disease is a fallacy and should be retired from medical thinking. Many diseases are caused by parasitic infection that the body probably never recognizes.

All around us, we daily see evidence that modern medicine produces unsatisfactory outcomes. Life expectancy drops, people are fatter than ever, diabetes mellitus is on the rise, and the health recommendations that are being made by the medical profession clearly don't work. If exercise were the answer to good health, then Jim Fixx, the famous runner and author, would not have dropped dead in the street while out for a jog; nor would fitness expert Bob Harper (*The Biggest Loser*) have died of a heart attack (he was resuscitated). If vegetarianism were the answer, then vegetarians would live significantly longer than omnivores. If avoiding alcohol and tobacco were the answer, then the clean-livers would outlive everyone else. None of these is true. Instead, we see that Jeanne Calment, a Frenchwoman and the officially-recognized oldest person who ever lived, was a smoker for 97 years, up until her death at age 122½, which is a cold slap in the face to doctors who tell patients to quit smoking. And we are reminded of the old expression, "There are more old drunks than old doctors."

There's a saying and a rule of logic that goes: The exception disproves the rule. Jeanne Calment teaches us that smoking tobacco cannot be inherently harmful to human health; it's just impossible that the oldest person who ever lived could be a smoker if smoking consistently produced a net loss in terms of health; a person can not deviate that far from the norm. Similarly, old, biased beliefs about alcoholic beverages are starting to change as people begin to recognize that the judicious use of alcohol can improve rather than destroy health. (Alcoholic beverages were originally invented to improve health, and Jeanne Calment drank wine daily.) Meanwhile, the new guiding principle in medicine seems to be that your body must be attacked with drugs that suppress your immune system, because your health problems are caused by your body attacking itself because it is too stupid to distinguish between self and other. Actually, though, your body knows what it is doing—it is just being overpowered by chronic infection and by your doctors' harmful treatments. It's actually doctors who are too stupid to distinguish between self and other, and the immunosuppressive drugs that they routinely prescribe are killing people and greatly inflating the cost of healthcare. To understand what is happening, a look at history will be helpful.

Before taking a look at history, though, I need you to understand that I have cured more than fifty allegedly incurable diseases, including diabetes mellitus, Parkinson's disease, multiple sclerosis, Lou Gehrig's disease, allergies, and many others—often poorly-defined diseases that have considerable overlap with other diseases. I have accomplished this by rejecting illogical assumptions and conclusions that "everyone knows."

Furthermore, it is obvious to me that the same organisms, approaches, and cures in these so-called "autoimmune" diseases will also apply to curing at least several cancers; the possibility that cancer is significantly different from the other 50+ "incurable" diseases that I've cured is insanely small. I will explain the etiology of the modern medical delusion by examining history. Let us first look at cancer, because this clearly illustrates the problem in modern medicine.

Cancer is a disease that has been recognized for more than 3,000 years. This is part of the problem. We are using definitions of disease that are in many cases so primitive that they are 3,000 years old—older than Christianity. In fact, when we look at the many diseases around us—so-called 'modern' diseases—we see that they actually are often hundreds or thousands of years old. Cancer is 3,000 or so years old, heart disease and kidney disease have been recognized for at least 2,000 years; diabetes mellitus has been recognized at least since Aretaeus of Cappadocia wrote about it around the year 140; schizophrenia (dementia praecox) has been recognized for at least a few hundred years; and, more recently, Parkinson's Disease was described in 1817 by James Parkinson; Hodgkins lymphoma was described in 1832 by Thomas Hodgkin, who was an anatomist rather than a physician; leukemia was discovered in 1845 by John Bennett; the grammatically incorrect multiple sclerosis was named in 1868 (possibly earlier), and Alzheimer's disease was identified in 1906 by Alois Alzheimer. A great many 'modern' diseases were "discovered" in the 19th century or thereabouts. Why is this important? The reason is this: the concept of microorganisms did not exist decisively until 1878, after Louis Pasteur proved it and made his presentation to the French Academy of Medicine (and it was not universally accepted for more than a decade afterward); therefore, the only explanations that early doctors could use to explain disease prior to 1878 were miasma, bad humors, or inherent dysfunction of the body. Two of these theories are no longer used. It's time to take the last of them down. Cancer was—and still is—considered to be an inherent dysfunction of the body. This idea is not only wrong in most cases, it is also extremely persistent.

On another front, the concept of autoimmune disease needs to be examined, too. Before the discovery of microorganisms, there was no need for a concept of autoimmunity; by 1903 the immune system had been theorized, and by 1912, John B. Murphy had developed the first immunosuppressive drug. By the 1930s, the Mayo Clinic had started using the newly discovered immunosuppressive hormone cortisol to treat rheumatoid arthritis. However, the first use of the actual term 'autoimmune' may have been by Nobel laureate Karl Landsteiner in 1939, who was investigating serum sickness at the time; later, in 1946, Arnold Rich of Johns Hopkins Hospital investigated periarteritis nodosa, which he associated with serum sickness and lupus erythematosis, calling them autoimmune diseases. Whoever originated the concept and the term, they are distinctly 20th century, and recent, but built on ancient ideas that should have been reevaluated, but have not been (inherent dysfunction of the body). It is especially important to note that the use of cortisol predates antibiotic drugs (penicillin-1941), thereby radically influencing the direction of medical thought—and not in a good way.

The concept of autoimmune disease seems to have captured people's imaginations, like the idea of Disneyland, and doctors never looked back. Soon these two diseases would merge as autoimmune disease became an explanation for cancer. There was at least one major problem with this, though: cortisol, while providing great initial relief of rheumatoid arthritis, turned out not only *not* to be a cure, but to actually make the disease worse over the long term. In fact, no matter how much immunosuppression doctors applied, the diseases just kept getting worse and worse until the patients died—often with horrible, horrible hunchbacks or scoliosis from the worsening effects of the immunosuppressive drugs. No amount of immunosuppressive drugs or radiation has ever cured a so-called "autoimmune" disease—yet those 100% consistent failures never caused doctors to reconsider their approach to these diseases; they just figured that they just needed a different version of the same treatment, and more of it. They never stopped to realize that they were using a defective model of disease.

So what do you call a disease that gets worse when you apply immunosuppression? . . . You call it an infection! There is no other term to use. This simple concept has been completely overlooked in the histories of cancer and "autoimmune" diseases. Doctors fail to distinguish between short-term (e.g. 5 years) and long-term (e.g. 40-120 years) success, so patients often appear to get better for five years and then die, and the doctor is never the wiser.

When John Bennett opened up a patient's blood vessels at autopsy in 1845, he had no way to explain the gelatinous mass that filled them (he called it a "suppuration of blood"); only four months later, Rudolph Virchow made the same observation in a different patient, but, doubting Bennett's explanation, named it "weisses blut," German for "white blood," which two years later he would change to leukemia (Greek for "white blood"). Although accurately descriptive, the term makes no effort to explain. They still didn't understand about infections (1865-1878 or so), so this disease finding was regarded as something innate. Today, we recognize that white blood cells proliferate in response to infection by microorganisms, yet despite our modern knowledge of this fact, the archaic belief in leukemia as an innate dysfunction still governs the approach by "modern" medicine to this disease, and results in tens of thousands of deaths every year as doctors try to kill the immune system instead of the infection—inexplicably.

It's also important to note that, while Pasteur proved the existence of microorganisms in certain diseases, decades passed before this fact was commonly accepted. The history of medicine contains few if any instances of new ideas being accepted immediately by doctors; mostly what happens is that medical innovators die in ignominy, as happened with Alexander Gordon, who advocated the washing of hands in 1795, or Ignaz Semmelweis, who was possibly driven insane by his colleagues' refusal to wash their hands 50+ years later. In the mid-1800s, Dr. Charles D. Meigs of Philadelphia probably killed hundreds if not thousands of women by his adamant refusal to wash his hands or to practice cleanliness. He would go straight from the autopsy room to the birthing room, proud of all of the blood and pus caked on his surgical clothes because of the experience that they denoted. All of this happened 2,000 or so years after the Indians got very good results in surgery via the use of cleanliness (as described in the Susrutha Samhita— there is some disagreement about its age), so the intransigence of doctors is legendary—2000-3000 years were required just to get doctors to wash their hands! That struggle was an evil reminiscent of when the great medical reformer Paracelsus (ca. 1493-1542) was probably murdered by assassins who were likely hired by the local physicians guild, who resented having their incompetence illustrated to the public.

Now when we discard the ridiculous delusion of "autoimmune" disease, we need to find another explanation for cancer. It should be obvious (but apparently it is not) that everyone has infections. Old people especially are severely debilitated by infections; this is why they die of "old age." I see pathognomonic signs of the ubiquitous disease on many peoples' legs, plainly on display in the summer, but doctors ignore these, making false attributions due to ignorance, arrogance, and stubbornness. The emaciation associated with old age is a sign of *disease*, not just old age! When singer George Michael dies at age 53, when actress Carrie Fisher dies at age 60, both after very similar years of struggle with psychological, pain, and addiction issues, it should be clear that we are all suffering to varying degrees from the result of chronic infection—chronic parasitic worm infection! This infection causes both emaciation and obesity—often simultaneously; cholesterol/fat is part of the body's response to chronic infection, to sequester infections and toxins. It causes at-times-terrible pain, psychological problems, insomnia, and degeneration of every organ in the body. What could be the cause of so many different diseases? . . . The answer is parasitic worms.

How could this happen, you might wonder? Aren't we supposed to be too clean? How could literally everyone have worms? Consider a particular prototypical worm infection: 1) It has been endemic in humans' environments for millennia. 2) It can penetrate the skin, including the skin of the feet. 3) It can penetrate any organ. 4) It is passed from the mother to the fetus in the womb, imitating genetic inheritance. 5) It is passed from the mother to the infant in breast milk (think: breast cancer). 6) Individual worms can live 75 years. 7) They can travel in the bloodstream. 8) They can reproduce sexually or asexually. 9) They are especially fond of the lymphatic system. 10) Infections are life-long. 11) No one has ever been fully cured of this infection. 12) Treating for this organism cures a wide variety of diseases, including Parkinson's disease, diabetes mellitus, multiple sclerosis, Lou Gehrig's disease, and many more. 13) Doctors adamantly refuse even to consider this disease, even when faced with pathognomonic proof. Physicians' irrationality reaches the point of insanity, just as with the case of Dr. Meigs and others in the 19th century. There is no way to avoid this brilliant, infectious parasite, especially when doctors refuse to acknowledge it or to help their patients with it. (In fact, when James Parkinson published his ground-breaking *Account of the Shaking Palsy* in 1817, one of the cases and its cure were attributed to worms and their treatment. Doctors have subsequently ignored this wisdom for 200 years, so far.) Science might change, but stupidity is forever.

Most doctors would almost certainly tell you that worms can't cause cancer. The answer to this is the classic statement that those who are ignorant of history are doomed to repeat it—he who knows only his own generation remains always a child. In 1926, Johannes Fibiger won the Nobel Prize for Physiology or Medicine by using parasitic worms to induce cancer, showing that worms *can* cause cancer (see chapter 15), but an even trickier aspect of this infection is that because the infection occurs in the womb, it might never even be recognized by the immune system. (It appears that worms are the sole cause of acne, suggesting some immune system recognition, but this inflammation could be due to the gram-negative bacteria that worms carry in their guts.) In 1960, Burnet and Medawar shared the Nobel Prize for Physiology or Medicine for theorizing and demonstrating that fetuses exposed to foreign antigens in the womb fail to recognize those foreign antigens after birth—therefore, life-long infections can be acquired in the womb. This explains virtually every current chronic disease! Unrecognized eukaryotic, parasitic worms proliferate without significant inhibition until the host dies. Much of what you think of as old age is reversible, thereby proving that "old age" is not due to age, but to infection, at least to some degree. This paradigm might also help to explain why women tend to live longer than men—they lose more worms via menstruation, childbirth, and breast-feeding, compared to men.

Many diseases that are believed to be genetic are actually inherited worm infections. For example, one tired old argument that is often made is that diabetes mellitus must be genetic because if one of two identical twins gets type 1 diabetes, the odds that the second twin will get diabetes are 50%. This argument is made by people who don't seem to understand the concept of identical (monozygotic) twins. If there's only a 50% chance that both identical twins will have a disease, then the disease is not genetic! In fact, the inheritance rate among identical twins is approximately the same for allergies as it is for schizophrenia (both are caused by worms)—50%, give or take. The concordance rate for genetic diseases in identical twins should be approximately 99.9%.

In the case of Acute Lymphocytic/Lymphoblastic Leukemia (ALL), a blood cancer, the defining characteristics include 1) very high numbers of defective white cells, 2) extreme bleeding, 3) super-white skin (in Caucasians). To explain these phenomena, let's start with the high numbers of white cells. The white cells are high due to some kind of infection (this is normal), but with an extreme copper deficiency as a result of inherited parasitic worms (increased demand on copper-zinc superoxide dismutase and P450 liver enzymes) that prevents the cells from reaching a mature level of development, thereby blocking a negative feedback loop, thereby leading to uncontrolled reproduction of cells to that blocked, immature level of development. A similar situation occurs in Acute Promyelocytic Leukemia, which has been cured using Vitamin D; the so-called cancerous cells go on to form normal macrophages, thus showing that the so-called cancer cells are not defective at all, but normal, though trapped at an immature level of development due to nutritional deficiencies presumably due to chronic infection and irritation. I feel certain that copper is necessary to form Vitamin D (although I have never seen this written anywhere) because copper is often used to hydroxylate molecules; so Vitamin D deficiency is consistent with copper deficiency (and melanin deficiency), showing parallels between the two diseases. Clearly, the problem in ALL does not lie solely in white cells, though, because white cells have nothing to do with bleeding—the usual cause of death in ALL.

Moving on to extreme bleeding, then, copper deficiency is known to lead to excessive bleeding by inhibiting Factor V (five) and Factor VIII, which both require copper. Copper is necessary for the development of various blood cells, including platelets. Finally, the extreme paleness of the skin of leukemic patients is due to melanin deficiency. Melanin gives color to hair, skin, and the iris of the eye, and its formation requires copper (and also Vitamin C, tyrosine, and some other things). This explains why my elderly father, who is increasingly frail, has gray hair, pale skin, and almost-white irises. Copper is also required to form the melanin that gives the black color to the substantia nigra, where dopamine is made, which is the area of the brain that is believed to be inhibited in Parkinson's Disease, copper also being required to form dopamine—which is deficient in Parkinson's—and the other catecholamine hormones. Copper is also required to make serotonin, which is also deficient in Parkinson's and schizophrenia. Worm-induced copper-and-other nutritional deficiencies explain everything perfectly, including autism (I cured a case of Asperger's syndrome), and people who undergo an aggressive treatment regimen report having the feeling of aging backwards, the change is so profound.

So if everyone has worms, and if aging is highly correlated with decreasing levels of copper, are there signs of chronic disease that are visible or is it something so arcane that virtually no one can understand it? There are many signs of increasing debility due to worms, but people have always just assumed that these things are not signs of disease, but are normal—because they are in fact normal! Parasitic worm infection is the most normal thing in the world! However, death is also normal! Normal isn't necessarily good for you. The signs of disease manifest at different ages, locations, and intensities for different people.

When I was young, I was scared of my grandmother. She was often kind of grouchy, and she had enormous, tree trunk legs, and she had giant loose flaps of skin that swung like wings from her triceps when she moved her arms, plus, her apartment had the dreaded "old people smell," a smell that seemed like decay. Parasitic worms explain it all, perfectly. I imagine that she must have suffered a great deal, just as innumerable people have throughout history, and often at the hands of incompetent doctors who tell patients that their problems are all in their heads, or that they just need a hysterectomy, or that they are just getting old, so deal with it. If that seems like harsh judgment of physicians, consider that these worms, or similar ones: 1) have been recognized for 140 years (1876—around the time of bacteria); 2) are inside every mammal, and 3) are described—albeit imperfectly—in every medical reference book. Yet doctors ignore them, just as they used to defiantly and irrationally refuse to wash their hands, their clothes, and their scalpels. How many centuries should it take for physicians to accept reality? Do we have to wait another 2000 years for doctors to believe the truth?

All Truth passes through three stages: First, it is ridiculed. Second, it is violently opposed. Third, it is accepted as being self-evident. —Arthur Schoepenhauer

Excellent evidence for worm infection is the example of addiction. Whether the addiction is to alcohol, tobacco, or heroin, it is explained perfectly by parasitic worms and the pain that they often cause, in most—perhaps not all—cases. (Some worms are reported to secrete immunosuppressive chemicals, perhaps cortisol, thereby reducing inflammation and pain.)

Most heroin addicts that I've seen have the emaciation characteristic of the starvation that is caused by worms, but without the fat production. The famous actor Phillip Seymour Hoffman was a heroin addict, but he had the classic obese presentation of parasitic worms—pale, lumpy skin; obesity; chronic pain; fatigue. I'm sure his doctors told him to diet and exercise— useless!! To be a good doctor, an understanding of chemistry is helpful: Alcohol is metabolized to acetic acid (basically vinegar) and inhibits worms. (Vinegar is used to treat diabetes mellitus—worms presumably love the pancreas so much because it is a very alkaline organ, plus it is close to the liver, where these infections start after descending the umbilical vein from the placenta.) Tobacco has nicotine, which is a powerful natural pesticide (good for the nerves). Caffeine is also a powerful natural pesticide. Addictive substances tend to be either natural pesticides or pain killers. (Think about it.) (Sugar is an exception; it is craved due to the starvation of parasitic worm disease.) Alcohol, teas, coffee, tobacco, marijuana, herbs, spices, vegetables—all have natural pesticides (unfortunately heroin does not seem to), many of which are the phytochemicals that are becoming increasingly popular as nutritional supplements. I believe that gluten is probably a natural pesticide (I've cured gluten allergy), albeit not one of the best for treatment purposes, and that is why it reacts so strongly in many people. Once the inhibitory drug effects against worms wear off, in a few hours (around eight hours for heroin), another dose of medicine is needed as the worms wake up and resume activity that can be quite painful if they are ravenously hungry.

The reason that smokers need to smoke immediately after waking up, and the reason that "sundowning" occurs in Alzheimer's disease, is because these worms are primarily nocturnal. Think of the numerous diseases that are worse at night (For about 30 years, I would wake up in more pain than when I went to bed—hint!): COPD, asthma, Alzheimer's, restless legs, IBS, IBD, GERD, rheumatoid arthritis (morning stiffness), insomnia, carpal tunnel syndrome, and many others—all worse at night! This is why having a smoke and a drink (alcoholic) before bed is very healthy. Paracelsus taught us that "Everything is poison. Only the dose makes a thing not a poison." This was simply the earliest statement of the therapeutic window, the Goldilocks effect: too little is useless, too much is harmful, and somewhere in between is an amount that is just right (helpful). (In *David and Goliath*, Malcolm Gladwell uses the inverted U-shaped curve to illustrate this phenomenon.)

This is why many centenarians, when asked by reporters what their secret to long life is, say "a glass of whisky (or brandy, or gin or whatever) every night"; the reporters usually laugh derisively, and say "isn't that cute, she thinks that whisky is the cause of her long life." Then the reporters, in the supreme arrogance of youth, ignore their elders' wisdom and lead much shorter lives. You can lead a horse to water, but you can't make him drink, and my dog doesn't like booze much, either—it's definitely an acquired taste.

This is why exercise is not the sole or even the main answer to long life: exercise doesn't cure the infection! (However, it does create lactic acid—which is presumably toxic to worms.) Don't exercise to get well; exercise to feel better! Vegetarianism is still good for long life, but it is more prevention than cure. Jeanne Calment, in case you were wondering, smoked an amount that has been described as equivalent to no more than two or three cigarettes per day, mostly in the evening. Cigarettes have more additives than pipe tobacco, so they may be very much less healthy. I would recommend natural tobacco in a pipe over cigarettes, and just a little before bed, if you're inclined to smoke. While smoking tobacco inhibits worms that lower copper levels, smoking also lowers copper levels, creating more complexity and increasing the risk of lung cancer (the vast majority of smokers never get lung cancer, though). Copper supplements and other nutritional supplements should be taken by smokers, especially, and they should be taken with Vitamin C. It's probably healthier to treat worms in ways other than by smoking, though. Interestingly, however, during the London Plague of 1665, Eton students were flogged if they failed to smoke before class in the morning, because of the belief that smoking inhibited plague, a belief that has biochemical support. Smoking of tobacco was used to treat a variety of diseases beginning around 1500, when tobacco was first brought to Europe. Native Americans had been using it for hundreds if not thousands of years before that, but they didn't keep records. The demonization of tobacco occurred after Austen Bradford Hill and Richard Doll did a study (1947-1951) in England that showed "an association between" smoking and cancer. Somehow, that "association between" morphed into "a cause of". It was largely (perhaps not 100%) a false attribution because they never realized that smoking, rather than being the initial event, is instead a secondary event, a response to an underlying infection. In 1962, The Royal College of Physicians declared that smoking caused cancer, and in 1964 the U.S. Surgeon General jumped on that train, too. Thus was the demonization of the tobacco herb off and running, and doctors never looked back.

Finally, when Matthew Baillie described the first recorded case of cirrhosis of the liver circa 1793, and attributed it to alcohol, he made a false attribution, an attribution error, just as many, many doctors have done before and after him, in reference to smoking and alcohol: he, like essentially all doctors after him, failed to see that alcohol was the treatment, not the disease! (It's not a perfect treatment, however.) There was really no way for him to know—history just wasn't ready yet. Alcohol consumption is a confounding variable in disease, and is usually/often a false attribution for it. I believe that the true cause of liver cirrhosis (and COPD, which I've reversed) is parasitic worm infection; but the therapeutic window must still be considered, and large quantities of alcohol are still not advisable. Remember Goldilocks. There are anti-worm drugs that are cheap and safe. One of them (ivermectin) won the Nobel Prize in 2015 for its inventors/ discoverers William C. Campbell and Satoshi Ōmura.

I hope that with this article we can begin to stop the enormous waste of lives and money caused by "modern" medicine so that we may apply our limited resources to worthy pursuits that are *not* gigantic wastes of time and money, such as "modern" medicine and the War on Cancer (and now the "Moon Shot for Cancer") have turned out to be, and which they continue to be—an enormous, misguided, incompetent drain on the public coffers— welfare for clueless scientists who have arrogantly ignored basic science and science history in their ill-considered rush to find glory in a search for genetic unicorns, while millions of people endure intolerable suffering (often as incompetent doctors withhold pain killers "for the patient's benefit, to prevent addiction"—"so-and-so has an 'addictive personality'", they say— No! It's an infection!!!) until they die in agony. Here's a clue: everyone has worms; worms cause severe pain and degeneration; everyone sooner or later needs either pain killers or—even better—competent medical treatment to kill their worms!!! Until they can get competent medical care, they should be able to get pain killers, and not have them withheld by the clueless, religion- controlled government! Lawmakers should stop trying to practice medicine, because they literally know nothing about it! If doctors don't have a clue about medicine, then legislators certainly don't.

Currently, doctors are certainly very little better than they were in the middle ages (surgeons and anesthesiologists excepted)—immunosuppressive drugs to treat "autoimmune" diseases and anti-cholesterol drugs are the modern versions of bloodletting, and should be a humiliating embarrassment to every doctor; these drugs do far more harm than good. Years from now people will shudder in horror at today's doctors just as we are appalled at the thought of the doctors who bled presidents and emperors, mothers and children, to death in an effort to save them, or who carried the same deadly germs from patient to patient for decades, creating mortality rates of up to 80%. Mental illnesses and addictions are treated using methods that are comparable to exorcism (methadone is not a treatment—it's a substitution, and a scam). Let us see if we can bring the world to an intelligent understanding of health and disease, one that's based on real chemistry and a real understanding of infectious disease, instead of the fanciful, short-sighted, worn-out, greedy, corrupted paradigm that is currently popular, whereby only the most expensive, patented pseudo-"treatments" and counterproductive drugs are considered. The old paradigm of "autoimmunity" is dead! Bury it and be done with it! Doctors are killing people and pets; let's fix this now. Don't wait 2,000 years; accept the truth now! For more information, please go to www.RobertSFarmerMD.com.

Talent hits a target that no one else can hit. Genius hits a target that no one else can see. —-Arthur Schoepenhauer

BIBLIOGRAPHY

- Porter, Roy; The Greatest Benefit to Mankind; W.W. Norton & Co., New York
- Kennedy, Michael; A Brief History of Disease, Science, & Medicine; Asklepiad Press, Mission Viejo, CA
- Mukherjee, Siddhartha; The Emperor of all Maladies; Scribner, New York
- Aptowicz, Cristen O'Keefe; Dr. Mutter's Marvels; Penguin Random House, New York
- DeVita, Vincent; The Death of Cancer; Sara Crichton Books

- Shils, Maurice E. et al; Modern Nutrition in Health and Disease, 10th ed.; Lippincott Williams & Wilkins
- Dobson, Mary; Disease; Quercus books, England
- Current Medical Diagnosis and Treatment; McGraw-Hill. My edition is from 2002; pick any year.
- The Merck Manual; Merck Research Laboratories, Whitehouse Station, N.J.
- Harrison's Principles of Internal Medicine; McGraw-Hill
- Brownstone, D.; Franck, I.; Timelines of the 20th Century; Little, Brown, and Company
- Gladwell, M.; David and Goliath; Back Bay Books and Little, Brown, and Company; N.Y., N.Y

Chapter 7

The Cure for Parkinson Disease, Multiple Sclerosis, ALS, Autism, etc.

I have heard him [William Harvey] say that, when his [revolutionary] book on blood circulation came out, his practice suffered and people called him mad. — John Aubrey

I want to tell you how I cured Parkinson Disease, but you will need to understand that the cause of Parkinson Disease is certainly the same as the cause for many other idiopathic neurological diseases, such as autism, multiple sclerosis, schizophrenia, Amyotrophic Lateral Sclerosis (ALS, Lou Gehrig's Disease), and Alzheimer Disease, to name only five of many. The most likely cause of these diseases was first discovered 136 years ago, in 1876,[1] but the full ramifications of this infection have not been fully appreciated until now. You are reading history in the making, right here. Let me start with some personal history.

I used to be extremely physically active, able to do pull-ups with one arm and literally run up 14,000 ft. peaks in Colorado, despite medical problems that had started years earlier, as an elementary-school student. It began with fatigue, diminished bowel control, and idiopathic stomach pains that frequently occurred in the mornings and often prevented me from going to school, but which usually resolved around noon. At eight or so years old, I was diagnosed with a gastric ulcer, and I took Maalox™ through most of my teenage years. In high school, I was a cross-country runner, but I had some chronic fatigue that prevented me from being outstanding. Also in high school, I was one of those kids who could never build muscle mass, no matter how much I lifted weights, and I had a recurring problem with shin splints. These were always very frustrating for me. I eventually became a rock climber and trained fanatically, but never seemed to achieve the standards that I thought that I deserved. Later, I took up peak-bagging, and I sometimes *ran* up the mountains in Colorado, once doing the two highest mountains in Colorado in the same day, a vertical gain and loss of about 10,000 vertical feet, each way. I was in better shape than most people could even *imagine* being in. Nevertheless, I always felt rather tired, even as a teenager, and more so in my twenties. Eventually, chronic bronchitis and emphysema (COPD) snuck up on me (even though I never smoked), and slowly, over a period of decades, I had to quit the activities that I had always enjoyed; despite being a health fanatic, I eventually deteriorated to the point where I would become winded going up a flight of stairs at sea level. It was very demoralizing. Sometime around then, I decided that in order to regain my health, I would have to go to medical school, because I didn't think that anyone else would cure my health problems for me.

Eventually, my health deteriorated to the point where I had osteoporosis that resulted in four broken legs for minor and inappropriate reasons, I was continuously exhausted, and I had continuous, debilitating pain in my back and spine, so that I could barely hold my head up for 15 minutes at a time. I gained a lot of weight, going up from 165 to 245 pounds despite my continuing efforts to be healthy. Eventually I suffered from what was probably early Alzheimer's Disease, and my hormonal system seemed to be shot, because I frequently suffered from low blood sugar, and I had to drink a lot of salt water to maintain my energy, indicating deficiencies of aldosterone[2] and glucagon.[3] Despite all of my problems, there was no disease that fit my symptoms all that well. My disease seemed like Addison Disease, but not perfectly; I also had acquired Fanconi's syndrome, losing nutrients from my kidneys.[4] I had tremors and twitches and balance problems and Restless Legs Syndrome along with scoliosis and a variety of other problems. I didn't really realize that I had Parkinson Disease during the time when it affected me most—I did recognize, though, that I seemed to have classic signs of

Multiple Sclerosis, such as bladder problems, incomplete voiding, premature bladder filling, visual deterioration, muscle twitches and tremors, etc. For several years, I thought that my problems, or at least my gastroesophageal reflux disease (GERD), were/was due to a fungal infection, as I clearly had a very obvious and very heavy fungal growth on my tongue (lab-verified fungus that many doctors were incapable of recognizing, by the way), and I took almost all of the anti-fungal drugs, up to voriconazole, but never Amphotericin B. These drugs helped some, but not that much. I was still dying, and I knew it.

I read every chapter, every page, in the main medical reference books, and eventually I noticed that my symptoms seemed to fit a parasitic worm infection, probably Strongyloidiasis--although, again, not perfectly. Infection with *Strongyloides stercoralis* causes a bewildering variety of physical problems, from gastrointestinal problems to respiratory problems to neurological problems to kidney problems and heart problems; here was an infectious organism that can infect any organ in the body![5,6] At that time, I, like every other doctor/medical student was still brainwashed into thinking that I, as a Caucasian American, could not get such a disease. Eventually, though, I realized that strongyloid worms are endemic in the environment,[7,8] and that American horse and dog owners are supposed to treat their pets for this infection virtually continuously,[9] so, in pure desperation, I considered that I might have a parasitic worm infection. In examining my family history, I also noticed that parasitic worms explained many diseases in my family history over the last 500 years (I have a famous ancestor, and my analysis skipped a few generations). I experimented with a variety of anthelminthic (anti-worm) drugs, and I thought they helped, but for a long time, I was not sure. The reference books said that one dose of medicine was good for six months, so for a long time I was afraid of taking too much and damaging my liver or kidneys. This prevented me from achieving therapeutic doses sooner, delaying my recovery and my findings, but anthelminthic medicines became the key.

One of the worst symptoms that I had was debilitating GERD (acid reflux disease). This turned into an even more debilitating gluten allergy (celiac disease), along with onset of more general allergies, as well (e.g. hayfever). I had to sleep sitting up in a chair, and I had to try to avoid gluten, which is ridiculous, because it is everywhere. Gluten wasn't the only food allergy, though; I seemed to be allergic to everything. It took some time, in part because I had been scared by the myths that the medical establishment has promulgated ("those drugs are dangerous"), but I eventually reduced my parasitic worm burden by more than 90% (estimated), thereby curing my Parkinson's Disease (which I diagnosed after the fact), Multiple Sclerosis, gluten allergy, general allergies, and numerous other diseases. I can now bench press more than 200 pounds and lift 300 pounds off the floor—more than ever—which was unimaginable just a few years ago.

I nearly died at age 49, the same age at which my mother died of "cancer." This suggested a pattern to me. As I looked around, I noticed more and more patterns. One of the observations that has always bothered me is that my mother deteriorated rapidly and died soon after she quit smoking tobacco and drinking alcohol, which was the opposite of what I had been taught should happen.

One time, I saw a news story about a guy who lived to be 103 years old. His niece said that the subject in question had been a very stubborn man who "'did just what he wanted to do," which included smoking tobacco and drinking coffee all day long. It was reported that he lived to be 103 despite these bad habits. That just didn't make sense in the medical model that we doctors have been taught; I had to think about that for a long time in an effort to make sense of it. Eventually, I realized that tobacco and coffee have strong anti-parasitic effects,[11-14] and that people who drink coffee live longer than people who don't.[11-15] I, who had no risk factors under the standard medical paradigm, had a proved parasitic worm infection, which was allegedly impossible; consequently, I finally realized that that old man, who used anthelminthic substances all day, every day, didn't live to be 103 *despite* his bad habits, he lived to be 103 *because* of them.

I've finally realized that much of what is commonly believed to be normal aging is in fact the deterioration caused by parasitic worm infection, and every person on earth has it. Think about it: we all had barefooted ancestors; the parasite can be transmitted in the womb;[16] it reproduces very slowly, by parthenogenesis,[17] so even one parasite can turn into billions; infections are lifelong;[18] and no one has ever been treated adequately for parasitic worms. The older a woman is, the more worms she will have at parturition, and the more worms she will pass to her children, so their infectious burden will increase dramatically more rapidly; the allergies caused by these worms are mediated by Immunoglobulin E (IgE), and are the cause of autism.[19] As women shift their childbearing to older ages, infantile infection loads and subsequent disability increase--autism, obesity, diabetes mellitus, schizophrenia, Alzheimer's, etc.[20]

Parasitic worms are eukaryotic organisms, and therefore are better able to hide from the immune system than can the prokaryotic organisms that we usually associate with infection. In fact, antibodies apparently don't work against these worms, nor do phagocytic cells; the only way that the body has to fight these infections is to slime/foam them with triglycerides and very low-density lipoproteins, producing atherosclerosis.[21] These substances form fat cells that secrete tumor necrosis factor-alpha (TNF-α) and interleukin-1-beta (1L-1β),[22] which have the effects of creating insulin resistance and hyperglycemia (by decreasing insulin secretion), respectively;[23,21] In other words, *the natural effect of the immune system is to cause atherosclerosis, obesity, and diabetes mellitus in response to parasitic infection!* (Obesity in the U.S. is primarily due to chronic infection rather than overeating or soda consumption, although excessive sugar is a harmful factor.) The IgE that the body secretes in response to parasitic infection stimulates the release of TNF-α and 1L-1β from the fat cells,[24] thus producing a wide range of so-called 'modern' diseases.

In my case, though, I suffered from hypoglycemia rather than hyperglycemia, but the mechanisms are essentially the same: parasitic worms are first implanted in the fetal liver, and from there it's a short jaunt just a few inches down the hepatic duct to the pancreas, where the parasites might damage either alpha or beta cells or neither, thereby causing either hypoglycemia or hyperglycemia or some other disease, depending on random factors. The hepatic duct empties into the duodenum (small intestine), and the liver also directly contacts another loop of the small intestine. The liver also contacts the colon directly, as well as the right kidney/adrenal gland and the gastroesophageal junction, which provides a convenient access to the pharynx, bronchi, and the lungs, not to mention the mouth, eyes, and brain. This, then, is the mechanism for celiac disease, irritable bowel syndrome, ulcerative colitis, COPD, and probably Crohn's disease, and several other diseases, too, including blindness (my eyeballs have shrunk as a result of anthelminthic therapy, indicating decreased intraocular pressure [i.e. in glaucoma]), allergies, adrenal fatigue, and mental illness, etc.

Some people delay onset of this disease by subconsciously using a variety of anti-parasitic substances such as alcohol, coffee, spices, tobacco, and marijuana. Hot chili peppers and garlic with olive oil seem to be very powerful, too. You have probably never noticed that vegetables such as carrots and celery have strong anti-parasitic properties, while, if you have ever bought corn fresh from a farmer's field, you would know that you should peel the top of the corn husk to check for worms before buying it. Therefore, I conclude that worms love corn, and therefore they must love high-fructose corn syrup, which is found in sodas, etc., these days, and is undoubtedly a main factor in modern obesity. Interestingly, while the original Coca-Cola™ was made from two strongly anti-parasitic plants, the coca plant and the kola nut from the kola tree, the modern version is loaded with high-fructose corn syrup that negates any beneficial effects that it presumably once had.

In my own case, as I realized that I had developed Parkinson's disease, the question of how to best speed my recovery became a matter of interest. Parkinson's disease isn't really all that complicated. It is usually said to be caused by a deficiency of dopamine. There are only two reactions involved in making dopamine, so how hard can it be? The amino acid tyrosine is converted to DOPA, then DOPA is converted to dopamine[25]—simple, right? So what can go wrong? We know that the effects of Parkinson's disease can be delayed for about five years by giving L-DOPA,[26] so the primary problem must be in the first, not the second, reaction, right? The second reaction works just fine, at least at first. So then, if we look at the first reaction, we see that it requires oxygen, vitamin C, tetrahydrobiopterin, copper (some people erroneously say iron), and tyrosinase.[25] I have no good idea about tetrahydrobiopterin, but I think I have enough oxygen, and I can certainly make sure that I have enough vitamin C, which I've always taken sporadically, anyway, and tyrosinase is something that I probably can't do anything about, so that leaves me with copper. My freakishly sun-sensitive skin already indicated to me that I might have a copper deficiency. So, I started taking copper. Actually, I had already known that I was deficient in copper, because I cut my hand once, and it bled every day for two months during my surgery rotation when I was washing my hands constantly, until I read somewhere that copper can help with healing; after taking copper, my hand stopped bleeding within 24 hours, and it never bled again.

I had been taking copper sporadically since then. Of course, I now know that copper is essential for lysyl oxidase, which cross-links collagen and elastin,[27] and for Factors V (five) and VIII, which are essential for blood clotting[28] Long story short, I cured my Parkinson's disease by treating for worms and replacing the lost minerals, especially copper. The question then arises: Why was I deficient in copper, and why are other Parkinson's patients deficient in copper? Copper, in addition to being essential for insulin function,[29,30] is also required by the immune system.[31] The function/mechanism is not fully known/described at this point, as far as I am aware, but the necessity for copper in the immune system has been proved.[31] So, then, a copper deficiency suggests a chronic infection. What in the world could cause such a chronic infection, that no one has recognized despite the presumed fact that literally everyone has it?

Have you ever noticed that everyone over the age of 40 and many people under the age of 40, and numerous children, have dark circles and bags under their eyes? I suppose that most people just think that having bruises under both eyes (bilateral suborbital ecchymoses) is a normal part of aging, but is it? Why would it be? Bruises are a sign of damage, trauma—is there ever a time when a spontaneous bruise is normal? What is it about aging that causes bruises as a normal function of getting older? The answer: nothing! You can argue about the definition of normal, but the evidence indicates that under-eye circles/bags are pathognomonic for parasitic worm infection! Did you ever notice that all of those schizophrenic mass murderers have the worst eye bags that you've ever seen? (And their faces often look like emaciated, parasite-ravaged skulls with just the thinnest layer of skin?) I say this because I have reversed these suborbital bruises using anthelminthic therapy. Gray hair,[32] pale skin,[32] wrinkles,[32] hormonal deficiencies,[32] autoimmune diseases,[32] bad teeth, sinus congestion, bad vision, sleep problems, spinal degeneration, sarcopenia (muscle loss), and other such factors are not signs of aging as much as they are signs of chronic parasitic infection and the copper deficiencies[32] that this infection causes. Actually, because everyone in the history of the world has had this infection, we do not even currently know what old age actually looks like! I had to stay out of the sun for years; for years, I would wear hats, sunglasses, and sunscreen every single day, or I would get horrible sunburn. Now, I don't need those, anymore. I tan fairly well, now (production of melanin, the substance that gives color to hair and skin, requires copper[33]) even though, as a blue-eyed blond, I generally tan worse than anyone other than albinos and some redheads. My vision is fairly good, despite still having worms in my eyes (called "floaters"—also, melanin [and therefore copper] is required for function of the eye[34]). Now, I am a very young-looking 52-year old man (the opposite of my previous appearance). This is proof that parasitic worms and the resultant copper deficiencies are the cause of Parkinson disease and Multiple Sclerosis, etc. I admit that I have not proved it in thousands of patients, but it will be proved beyond any doubt, I am certain—if only I can get the word out so other people can verify my findings. My discovery has cured type-2 diabetes mellitus and Asperger's syndrome in others, along with many diseases that I have had. Occam's Razor states that the simplest explanation is the most likely, and there are no explanations that could be simpler and more logical—it's an infection! It's probably just cultural arrogance that has prevented us from recognizing this earlier.

In analyzing neurological diseases such as Parkinson's disease, Multiple Sclerosis, and ALS, it's also useful to note that copper is essential in a variety of important and essential functions, such as in forming the myelin sheath that covers large nerves,[35] so it therefore seems that copper should be an obvious consideration in any neurological disease. It is this sheath that suffers the multiple scleroses (injuries, scars) that define Multiple Sclerosis. Therefore, Multiple Sclerosis is almost by definition a state of copper deficiency, otherwise the scleroses would heal, probably using lysyl oxidase, which requires copper to cross-link collagen and elastin. Copper is also essential in the formation of the red blood cell, being responsible for transporting the iron into the hemoglobin using ceruloplasmin.[36] Cases of apparent iron-deficiency anemia that are unresponsive to iron therapy are usually due to copper deficiency.[37] Also, copper is essential for cytochrome c oxidase, which is an essential part of the electron transport chain, and so is essential for life.[38]

Interestingly, doctors pretty much never consider copper. One reason for this is that copper levels in the blood stay fairly constant despite wide fluctuations in the body's copper content, the liver apparently being very good at maintaining necessary levels in critical areas.[39] Therefore, the best way to measure copper levels has been with a liver biopsy, but everyone is reluctant to perform such risky and painful invasive procedures.

Also, convincing proof of the importance of copper and its essentiality in the immune system are relatively recent findings (since 1964).[40] The main deficiency in the understanding of these many diseases seems to lie in the refusal of people to consider parasitic worm disease as a possibility; many people consider the suggestion that they have a parasitic (worm) infection to be an insult, rather than merely an objective observation or explanation. My findings, then, are the result of the application of pure reason, free of any prejudice. Certainly, there can be few other possibilities for explaining a disease that has probably been passed from generation to generation for tens of thousands of years and is treatable with anthelminthic drugs, and which is also present in dogs,[41] cats,[42] and horses,[43] as well. Strongyloidiasis has a method of inheritance that explains many so-called idiopathic "genetic diseases with variable penetrance." And while there has been speculation that copper may be a causative factor in Alzheimer disease,[44] it is important to realize that copper is used by the body to fight parasitic infections,[45] and also by sheep owners treating their sheep,[46] and fish owners treating their fish,[47] thereby suggesting that researchers have confused cause and effect, as often seems to happen (such as with tobacco[48] and alcohol[49-51]), because of preconceived, primitive notions of disease that are just wrong.

These parasitic worms, most probably *Strongyloides stercoralis,* will be found to cause many diseases, perhaps more than 100 of them. This organism might even disprove the existence of autoimmune diseases entirely, diseases of which no one has ever been cured (thereby being an ineffective model of disease), as far as I am aware. Certainly, we now have the tools to reduce a great deal of unproductive and even harmful medical treatments and thereby reduce a great deal of suffering and waste of medical resources. For more information, please go to www.RobertSFarmerMD.com.

Since this article was written, I have reversed and apparently cured one case of Lou Gehrig's disease (ALS), giving me a 100% success rate at curing this disease.

REFERENCES

1   http://en.wikipedia.org/wiki/Strongyloidiasis. Accessed March 21, 2013.
2   *Merck Manual. Adrenal Disorders.* Beers MH, Berkow R, eds. Merck Research Laboratories, Whitehouse Station, NJ, 1999. 101.

3   *Merck Manual. Disorders of Carbohydrate Metabolism.* Beers MH, Berkow R, eds. Merck Research Laboratories, Whitehouse Station, NJ, 1999. 105.

4   *Merck Manual. Congenital Anomalies.* Beers MH, Berkow R, eds. Merck Research Laboratories, Whitehouse Station, NJ, 1999. 2231.

5   Tierney LM, McPhee SJ, Papadakis MA. *Current Medical Diagnosis and Treatment 2002, 4th edition.* Lange Medical Books/McGraw-Hill, New York, 2002. 1526-1528.

6   http://emedicine.medscape.com/article/229312-over-view, accessed March 26, 2013.

7   http://animaldiversity.ummz.umich,edulaccounts/Strongyloides_stercoralis/. Accessed March 21, 2013.

8   http://tmcr.usuhs.edu/tmcr/chapter13/intro.htm. Accessed March 21, 2013. Uniformed Services University of the Health Sciences, Tropical Medicine Central Resource.

9   http://www.greenbeltvet.com/equinedeworm.asp. Accessed March 21, 2013.

10 Lacy CF, Armstrong LL, Goldman MP, et al. Ivermectin. *Drug Information Handbook, 7th Ed.* Lexi-Comp, American Pharmaceutical Association, Hudson, Ohio. 1999. 644-645.

11 http://en.wikipedia.org/wikiTobacco_smoking. Accessed March 21, 2013.

12 http://en.wikipedia.org/wiki/Nicotine. Accessed March 21, 2013.

13 http://en.wikipedia.org/wiki/Coffee. Accessed March 21, 2013.

14 http://en.wikipedia.org/wiki/Caffeine. Accessed March 21, 2013.

15 Freedman, N. D.; Park, Y.; Abnet, C. C.; Hollenbeck, A. R.; Sinha, R. (2012). "Association of Coffee Drinking with Total and Cause-Specific Mortality." *New England Journal of Medicine* **366** (20): 1891-1904. doi:10.1056/NEJMoa1112010. PMC 3439152. PMID 22591295.//www.ncbi.nlm.nih.gov/pmc/articles/PMC3439152/.

16 http://books.google.com/books?id=uPKefgWTq3IC&pg=PT146&lpg=PT146&dq=strongyloidiasis+%2B+intrauterine&source=bl&ots=6fl9UUOtus&sig=Rq4xtj90G3yBB3nc3uj81LOcscO&hl=en&sa=X&ei=09dLUbINuPk4AOrpoD4Bg&sqi=2&ved=OCDEQ6AEwAQ, accessed March 22, 2013. (Bennett BT, Abee CR, Henrickson R. *Nonhuman Primates in Biomedical Research, Diseases.* Academic Press. London. 1998. 133.)

17 Ash LR, Orihel TC. *Intestinal Helminths.* In: Balows A, Hausler WJ, Herrmann KL, et al, eds. *Manual of Clinical Microbiology, Fifth Edition.* American Society for Microbiology, Washington, D.C. 786.

18 Tierney LM, McPhee SJ, Papadakis MA. *Current Medical Diagnosis and Treatment 2002, 41st edition.* Lange Medical Books/McGraw-Hill, New York, 2002. 1527.

19 Farmer RS. RobertSFarmerMD.com. *Autism.* accessed March 22, 2013.

20 Farmer RS. RobertSFarmerMD.com. (Various articles) accessed March 22, 2013.

21 Cannon JG. Cytokines and Eicosanoids. In: Shils M, Shike M, Ross AC, et al, eds. *Modern Nutrition in Health and Disease, 10th edition.* Lippincott Williams & Wilkins, Philadelphia, 2005. 663.

22 Brodsky IG. Hormones and Growth Factors. In: Shils M, Shike M, Ross AC, et al, eds. *Modern Nutrition in Health and Disease, 10th edition.* Lippincott Williams & Wilkins, Philadelphia, 2005.640.

23 Brodsky IG. Hormones and Growth Factors. In: Shils M, Shike M, Ross AC, et al, eds. *Modern Nutrition in Health and Disease, 10th edition.* Lippincott Williams & Wilkins, Philadelphia, 2005. 649.

24 Cannon JG. Cytokines and Eicosanoids. In: Shils M, Shike M, Ross AC, et al, eds. *Modern Nutrition in Health and Disease, 10th edition.* Lippincott Williams & Wilkins, Philadelphia, 2005. 657.

25 Champe PC, Harvey RA. *Biochemistry,* 2nd edition. J.B. Lippincott Co., Philadelphia, 1994. 266-7.

26 Beers MR, Berkow R, eds. Disorders of Movement. *Merck Manual.* Merck Research Laboratories, Whitehouse Station, NJ, 1999. 1468.

27 Turnlund JR. Copper. In: Shils M, Shike M, Ross AC, et al, eds. *Modern Nutrition in Health and Disease, 10th edition.* Lippincott Williams & Wilkins, Philadelphia, 2005. 287.

28 Turnlund JR. Copper. In: Shils M, Shike M, Ross AC, et al, eds. *Modern Nutrition in Health and Disease, 10th edition.* Lippincott Williams & Wilkins, Philadelphia, 2005. 288.

29 Fields M. Nutritional Factors Adversely Influencing the Glucose/Insulin System; *Journal of the American College of Nutrition,* August, 1998, *http://www.jacn.org/ contentlI7/4/317.full,* accessed Nov. 17, 2012.

30 Fields M, Reiser S, Smith JC Jr. *Effect of copper or insulin in diabetic copper-deficient rats.* US National Library of Medicine, National Institutes of Health. http://www. ncbi.nlm.nih.gov/pubmed/6344092, accessed Nov 17, 2012.

31 Turnlund JR. Copper. In: Shils M, Shike M, Ross AC, et al, eds. *Modern Nutrition in Health and Disease, 10th edition.* Lippincott Williams & Wilkins, Philadelphia, 2005. 289.

32 Turnlund JR. Copper. In: Shils M, Shike M, Ross AC, et al, eds. *Modern Nutrition in Health and Disease, 10th edition.* Lippincott Williams & Wilkins, Philadelphia, 2005. 286-299.

33 Turnlund JR. Copper. In: Shils M, Shike M, Ross AC, et al, eds. *Modern Nutrition in Health and Disease, 10th edition.* Lippincott Williams & Wilkins, Philadelphia, 2005. 289.

34 http://www.ncbLn1m.nih.gov/pmc/articlesIPMC32708911, accessed March 22, 2013.

35 Turnlund JR. Copper. In: Shils M, Shike M, Ross AC, et al, eds. *Modern Nutrition inHealth and Disease, 10th edition.* Lippincott Williams & Wilkins, Philadelphia, 2005. 289.

36 Turnlund JR. Copper. In: Shils M, Shike M, Ross AC, et al, eds. *Modern Nutrition in Health and Disease, 10th edition.* Lippincott Williams & Wilkins, Philadelphia, 2005. 287.

37 Turnlund JR. Copper. In: Shils M, Shike M, Ross AC, et al, eds. *Modern Nutrition in Health and Disease, 10th edition.* Lippincott Williams & Wilkins, Philadelphia, 2005. 286.

38 Turnlund JR. Copper. In: Shils M, Shike M, Ross AC, et al, eds. *Modern Nutrition in. Health and Disease, 10th edition.* Lippincott Williams & Wilkins, Philadelphia, 2005. 288.

39 Turnlund JR. Copper. In: Shils M, Shike M, Ross AC, et al, eds. *Modern Nutrition in Health and Disease, 10th edition.* Lippincott Williams & Wilkins, Philadelphia, 2005. 292.

40 Turnlund JR. Copper. In: Shils M, Shike M, Ross AC, et al, eds. *Modern Nutrition in Health and Disease, 10th edition.* Lippincott Williams & Wilkins, Philadelphia, 2005. 286.

41 http://connection.ebscohost.comlc/articles/34925319/strongyloides-stercoralis-infectionfinnish-kennel, accessed March 22,2013.

42 Farmer RS. *Treatment Notes* 2. RobertSFarmerMD.com, accessed March 22, 2013.

43 http://www.extension.org/pages/10281/strongyloides-threadworm-in-horses, accessed March 26,2013.

44 Turnlund JR. Copper. In: Shils M, Shike M, Ross AC, et al, eds. *Modern Nutrition in Health and Disease, 10th edition.* Lippincott Williams & Wilkins, Philadelphia, 2005. 296.

45 Turnlund JR. Copper. In: Shils M, Shike M, Ross AC, et al, eds. *Modern Nutrition in Health and Disease, 10th edition.* Lippincott Williams & Wilkins, Philadelphia, 2005. 297.

46 http://www.scsrpc.org/SCSRPC/Publications/part5.htm. accessed March 22,2013.

47 http://www.merckmanuals.comlvetlexotic_and_laboratory _animals/fish/parasitic_diseases_of_fish.html, accessed March 21, 2013.

48 http://en.wikipedia.org/wiki/Health_benefits_of_smoking, accessed March 22, 2013.

49 http://en.wikipedia.org/wiki/Alcohol_and_health, accessed March 26,2013.

50 http://en.wikipedia.org/wiki/Alcoholic_beverage, accessed March 26,2013.

51 http://en.wikipedia.org/wiki/Wine, accessed March 26, 2013.

Chapter 8

Revealing a Fundamental Cause of Lung Disease

Medicine is a science which hath been (as we have said) more professed than laboured, and yet more laboured than advanced; the labour having been, in my judgement, rather in circle than in progression. For I find much iteration, but small addition. — Francis Bacon

Summary

Lung diseases such as COPD, asthma, and cancer cannot be adequately explained or treated using the standard autoimmune/genetic disease paradigm, and treatment outcomes using this paradigm have been disappointing. However, these diseases are explained perfectly as parasitic infections, and can be and have been cured as such. The author claims that an endemic parasitic worm infection has been passed in humans (and other mammals) from mothers to fetuses from our earliest ancestors over tens of thousands of years and essentially constitutes normal, albeit destructive, fauna. These parasites cause many "idiopathic" diseases, among which COPD has been shown to respond to anthelminthic therapy, and has been reversed (cured) successfully. This parasitic infection paradigm is a much better explanation of diseases that are currently assumed to be genetic with variable penetrance or idiopathic. Deficiency of copper and other minerals as a result of chronic infection is a complicating factor in these diseases and should be addressed as part of therapy. Treating these many diseases appropriately as parasitic worm infections will dramatically lower health care costs and improve the quality of care as well as the quality of life for billions of people around the world.

What is the key question?
- What is the cause of COPD, asthma, and lung cancer?

What is the bottom line?
- COPD, asthma, and probably lung cancer are caused by parasitic worms that can be treated, leading to cure.

Why read on?
- Traditional thinking about these diseases is wrong, and needs to be changed. These lung diseases are curable, and relatively easily, at relatively low cost, but primarily over a period of years.

Introduction

Chronic obstructive pulmonary disease (COPD), asthma, and lung cancer are some of the most common and debilitating diseases on earth, with lung cancer being the most common form of cancer,[1] and with COPD being projected to be the fourth leading cause of death worldwide by 2030.[2] The traditional view of causation of these diseases is generally presumed to be genetic, with exacerbation by environmental activities such as smoking tobacco.[3-5] I believe that this view is mostly wrong, being based on prejudice, primitive beliefs, and scientific attribution error.

In examining the etiology of lung disease, an examination of the presumed causation is appropriate. Most experts attribute the cause of COPD to smoking tobacco, yet up to 55% of cases of COPD occur among people who never smoked;[6] attributing the cause of a disease to an environmental exposure that half of patients don't have seems to be indefensible at best and very bad science, in general. Additionally, only 10-20% of heavy smokers develop lung cancer, while only 15-50% of chronic smokers develop COPD.[7] Yet, there can be little doubt that smoking is found at a higher rate in people with lung disease than in the general population (approximately 45-80%

versus approximately 20%).[8] However, the high rate of never-smokers who develop COPD eliminates smoking as a likely causative factor. Still, though, many people with lung disease smoke, and many people who don't smoke get COPD. Why?

In the United States, there are currently some anti-smoking television commercials featuring people who smoked or who were exposed only to second-hand smoke, who developed lung cancer and/or COPD, and various other diseases. These people have presumably eliminated the alleged initiating event that caused their cancers/COPD, and yet they appear not to have developed improved health as a result. If smoking is really the cause of their conditions, and they stop smoking, wouldn't they be expected to heal? Wouldn't their health improve after a while? If the stimulus for disease is removed, shouldn't the body recover? Yet, the health of these patients remains bad. Why?

In asthma, an allergic reaction causes airway constriction. This allergic reaction is modulated by Immunoglobulin E (IgE),[9] a specific antibody. This antibody apparently appears for no reason, in an alleged "autoimmune" manner, and often in young patients. Why?

The answer to these questions is "parasitic worms," most likely *Strongyloides stercoralis*, a microscopic parasitic roundworm—or something very similar to it, perhaps *Dirofilaria* or *Brugia*, *Wuchereria*, or something as yet undiscovered; however, the hyperinfection syndrome of Strongyloidiasis fits the criteria perfectly.[10,11] While I have seen these worms, I am unqualified to offer a definitive identification in these diseases.

In reading the following discussion, I urge you to begin with a tabula rasa—a clean mind—and concentrate only on true scientific facts while rejecting the many prejudices/scientific errors with which we have been brainwashed over more than 100 years. I am going to confront and defy some of your most fundamental beliefs, and it will be painful and probably disorienting for you, but the truth is the truth, and I hope you will feel refreshed by the new clarity that you will have after you verify my findings in your own lives.

Discussion

Most people say that there is no reason to suspect a parasitic worm infection in first-world residents. I say that there is absolutely no reason to believe that anyone on earth does *not* have this infection. The argument against endemic parasitic infection is usually that these worms affect only people who have exposure to dirt or sewage, and we first-world people are therefore not susceptible, especially if we live in New York City or someplace that has no dirt or plants, unless we travel to a park or a farm. This is overly simplistic. The problems with this argument include the fact that these worms infect all organs, including bones and the uterus,[12] and that they can be (and always are) transmitted from the mother to the fetus in the womb and to the infant in breast milk.[13,14] Eventually, everyone shows signs of parasitic infection, but these signs have always been assumed to be (mistaken for) 'just old age.' It's a case of false attribution.

In medical school, we were taught that there are only two causes of increased IgE in the body: 1) allergies and autoimmune diseases, and 2) parasites. The problems with this claim include the fact that no one has ever been cured of an allergy or an autoimmune disease, thereby suggesting that the allergy/autoimmune disease model is defective and fallacious, a relic of a primitive, pre-technological era, like the native American explanation for the formation of Devil's Tower (involving a native princess, rock gods, a giant bear, and the instantaneous magical growth of a rock into the tower). Another problem with this allegation is that while treating allergies and autoimmune diseases never results in a cure of any disease, the *adequate* treatment of parasitic worm infection results in cures of allergies and autoimmune diseases 100% of the time, in my limited experience, suggesting not co-occurrence, but rather a cause-and-effect relationship. Diseases that I have successfully cured by treating for parasitic worm infection include neurological (e.g. Parkinson's, Asperger's, multiple sclerosis), "autoimmune" (e.g. allergies, celiac disease, irritable bowel syndrome, gastroesophageal reflux disease [GERD], diabetes mellitus), musculoskeletal (e.g. scoliosis, spine degeneration, chronic pain, joint diseases), pulmonary (e.g. COPD), and a variety of other diseases, all of which have never previously been adequately or accurately explained. Remember that these parasites infect any organ in the body, thus they can cause a bewildering variety of diseases, and they are very fond of both nerves and the lymphatic system. Also, they multiply so slowly, by parthenogenesis,[15] that host reproduction is often not impeded, and the infection becomes mistaken for old age, making it perhaps the world's most perfect destructive parasite, having gone essentially unnoticed for millennia. (*Strongyloides stercoralis* was discovered in 1876, and has been largely ignored ever since, even though it is endemic worldwide.) Astonishingly, we don't even know what parasite-free old age really looks like, because we have never seen it—it has never existed, ever.

Occam's Razor states that the simplest explanation is the most likely, if it explains the phenomena in question; this parasitic explanation explains many diseases perfectly, and the resultant cures prove it. Unfortunately, people demonstrate great unwillingness to admit that they might have a parasitic (i.e. worm) infection; this contention that people have parasitic worms usually incites a visceral response in people similar to calling their mother a drunken syphilitic whore, but this response is really quite inappropriate. Parasitic worm infection is not an insult or a moral defect; it is merely an infection of no moral significance—being essentially of normal-but-destructive fauna—and we need to start regarding it as such.

One of the apparently-unrecognized complications of this chronic parasitic infection is copper deficiency, which is an important key to curing Parkinson disease (my discovery), as an aside. Signs of this deficiency are blatantly obvious, but no one is aware of them, or people fail to make the association. They include: 1) gray hair (melanin inhibition) 2) pale skin with sunlight sensitivity (melanin inhibition) 3) loose, wrinkled skin (lysyl oxidase inhibition) and 4) easy bruising/bleeding (Factors V and VIII inhibition).[16] Parasites not only destroy the integrity of the skin and other tissues directly (causing flabby arms, jowls, etc. in the elderly), but by inhibiting lysyl oxidase, they also prevent its repair, which returns us to the people in the television commercials. The reason that they don't significantly improve after being treated for cancer is because their parasitic infections are progressing unimpeded (unless they got full-body irradiation for a bone marrow transplant), and their healing mechanisms are still impaired by chronic copper deficiency, which is in turn caused by chronic parasitic infection, combining direct and indirect effects, although many anti-cancer drugs undoubtedly inhibit worms, but at the cost of a Pyrrhic victory. These patients have been misdiagnosed, or, at the very least, the diagnosis did not go deeply enough, down to the true fundamental cause. Copper deficiency can be regarded as a potential critical factor in the development of many cancers because copper has been shown to be essential to blood cell formation and immune function.[17] There's a good reason that cancer patients often look pale and sickly—it's melanin deficiency due to copper deficiency. The aforementioned signs of copper deficiency are always mistaken for normal aging, yet isn't it strange that "old age" appears at so many different ages? Some people die of "old age" in their 40s, while others live to be well over 100 years of age before succumbing. Why?

I once saw a television news story about a man who lived to be 103 years old despite smoking tobacco and drinking coffee all day long, every day. It was remarkable that he lived that long despite these bad habits, the story went; it went against everything that we have been taught. I thought about that for a long time—it just didn't make any sense—before I realized that he didn't live that long *despite* his bad habits, he lived that long *because* of them. He did not defy the laws of physics, after all. Coffee and tobacco inhibit parasitic worms due to natural plant-made pesticides such as caffeine, nicotine, and acetaldehyde, etc.[18-22] This is why smoking tobacco has been falsely accused of causing most lung cancer; people subconsciously adopt smoking to treat the anxiety/neurological disease that parasitic worm disease often causes. In addition to causing anxiety, this infection also damages the lungs and other organs, including breasts. People smoke to treat their chronic, anxiety-inducing, undiagnosed parasitic infections, and because of misattribution (attribution error) due to prejudice, this smoking is then blamed for being causative of cancer and COPD when in fact it should be identified as being merely an associated, even a potentially beneficial, finding, one that is *induced by* disease rather than being *causative of* disease. While there can be no doubt that tobacco in sufficient doses can cause diseases including cancer—such as with chewing tobacco and oral cancer—"the dose makes the poison."[23] The allegation that smoking usually or even often causes cancer is overstated; smoking is a subconsciously-motivated response to disease, rather than *being* the cause of disease, although I'm sure that the interactions are sometimes complex and not straightforward. Just as arsenic is an essential nutrient at low doses and a deadly poison at high doses,[24] the relatively low levels of tobacco smoke inhalation are probably usually insufficient to cause cancer, instead distracting from the true cause of these diseases—parasitic worms. It's a case of false attribution; smoking is an unrecognized confounding variable. It appears that smoking damages parasites more than it damages the host, creating a beneficial effect that is nevertheless dirty and disgusting, with significant collateral damage as side effects. While smoking undoubtedly helps to treat parasitic infections, other treatments have far fewer negative side effects, so smoking as therapy can certainly not be recommended as treatment, unless one can't find a doctor to make the proper diagnosis and initiate the proper therapy, as is currently the case virtually everywhere. Smoking would then be regarded as the treatment of last resort, which is how it is currently being used, but without conscious awareness of the purpose or mechanism of action. If you want to cure smoking, cure parasitic infections first.

Having discussed COPD and cancer, let us now look at asthma. The body's first response to parasitic infection is to secrete IgE to fight the infection.[25] This is obviously associated with asthma, as IgE is regarded as one of the primary mediators of asthma. Less obviously, IgE eventually stops working against these parasites, if it ever works at all, for reasons that are unclear but which may be related to copper deficiency. As the infection progresses, phagocytic cells apparently become ineffective against these parasites at some point—strongyloidiasis is a life-long infection—so the body then attempts to control the infection by secreting fats, specifically triglycerides and very-low-density lipoproteins to foam the parasites. These form fat or foam cells that can cause atherosclerosis.[26] However, many people don't know that fat cells secrete cytokines such as Tumor Necrosis Factor alpha (TNF-α) and interleukin 1-beta (IL1-β),[27] which cause insulin resistance and decreased insulin secretion/increased hyperglycemia, respectively.[28, 29] This is important because IgE release is a stimulus for fat cells to release these two cytokines,[30] thereby establishing direct, consistent, causative sequential links via the immune system from parasitic infection, asthma/allergies/autoimmune diseases, atherosclerosis, obesity, diabetes mellitus, and death, in that sequence, with occasional detours to other diseases, such as lung disease, mental illness (many), dementia, heart disease, kidney disease, etc.

Having noted earlier that nicotine, an ingredient in tobacco smoke, is effective against parasitic worms, one might wonder if it might help in asthma. While many asthmatic patients smoke, in addition to being an acetylcholinesterase inhibitor and therefore an antiparasitic drug, nicotine is also a vasoconstrictor,[31] a counterproductive quality, thereby making it unpredictable and not especially useful as a treatment for asthma. Other methods are more effective and safer. Also, the advent of marijuana use to treat many diseases is undoubtedly due to its anti-parasitic effects, too.

The reason that immunosuppressive drugs are successful in these idiopathic lung diseases in small doses over the short term but fail to cure them in the long term is because they relieve the immediate inflammation but do nothing to inhibit the underlying infection, certainly even exacerbating the infection by inhibiting the immune system—an infection that cannot be treated by the usual, antibacterial antibiotic drugs. Doctors never consider parasitic worms (let's face it, most doctors never read the chapter on worms), so the diseases progress. Additionally, doctors tell people not to smoke tobacco or to use alcohol (ethanol), two anti-parasitic substances that slow the infection; ethanol is especially helpful in small-to-moderate doses, being toxic to numerous microorganisms (due to metabolism to acetaldehyde and acetic acid) while being relatively safe for humans, so saying to avoid it is clearly bad advice. People who drink up to three glasses of wine per day live longer than nondrinkers.[32] Moderation is key. Again, the dose makes the poison. Interestingly, these two substances are often prohibited and/ or usuriously taxed by governments—arguably, an impediment to public health. This raises questions of social justice, as we often fund many and varied programs via taxation on the backs of the poorest and most disease-debilitated citizens, who literally need these substances to survive.

It is unclear why no one has ever noticed the high/universal prevalence of these microscopic worms before, but one hypothesis is that because both worms and humans have eukaryotic cells, seeing the worms may be very difficult, as opposed to seeing prokaryotic cells, which would show more contrast against a background of human cells. It is odd, though, because these worms are often visible as floaters in the eyes, which many people have, so one marvels at how no one has noticed them or correctly identified them until now. Also, treating this infection will often require years of intermittently intensive therapy, and not simply one dose of mebendazole, which has often been used to treat pinworms (enterobiasis, probably usually unverified) in children,[33] but which is certainly dramatically inadequate for threadworm (strongyloidiasis)[34] or whatever worm this turns out to be; perhaps the modern short attention span is why doctors have not considered this infection before—this disease requires years of focused attention.

Because it appears that everyone has these parasites, they can also be presumed to be important in the treatment of cystic fibrosis (CF), which exhibits many signs and symptoms of parasitic worm infection, i.e. diseases of the lungs, pancreas, and gastrointestinal tract. When the bugs are first implanted in the fetus in utero, they come down the umbilical vein from the placenta to implant directly into the fetal liver, which is ideally located as a staging point for many diseases of the lungs, pancreas, GI tract, back, etc. Additionally, the chloride transport problem in CF can be presumed to interfere with the formation of hydrochloric acid, which is important in the stomach for digestion and probably also immune response to microorganisms such as parasitic worms both in the stomach and probably elsewhere. It is standard practice for CF patients to use digestive enzyme supplements to combat malabsorption. Even in parasitic worm infections that are not complicated by CF, decreased acidity (achlorhydria) is a recognized problem,[35] and the oral administration of apple cider vinegar is helpful, and is toxic to parasites, as are some alcoholic beverages, which are eventually metabolized to acetic acid (essentially vinegar), thus raising acidity. The current fad of increasing alkalinity of the diet by avoiding acidic foods is therefore completely wrong and counterproductive, in most cases, despite the fact that it may provide temporary relief of stomach discomfort. In fact, treating these worms properly will temporarily increase esophagus/ stomach discomfort due to the Jarisch-Herxheimer (JH) reaction from the toxins released by the dying parasites. This is why GERD/achalasia gets worse—people unwittingly use treatments that reduce the unpleasant Jarisch-Herxheimer reaction (e.g. heartburn), thus allowing the parasites to multiply and thrive. This JH reaction is a major reason why ethanol, garlic, chili peppers, etc., which are toxic to parasites, produce pain as they go down the esophagus, but not when they are rubbed on unbroken skin—there are more worms in the esophagus (the major cause of esophagitis), which contacts the liver (the staging area) directly. I now believe that the common hangover (malaise due to consumption of ethanol) is primarily or largely a JH reaction due to these parasites.

Finally, in addition to the classic signs of copper deficiency mentioned earlier, there is one very-easy-to-observe sign that appears to be pathognomonic for parasitic worm disease. That sign is circles/bags/bruises under the eyes (bilateral suborbital ecchymoses), which are ubiquitous above a certain age and have previously been documented in *Dirofilaria* infections.[36] While the old wives tale that these bags are caused by lack of sleep persists, there is no truth to this myth—again, there is an association, but it's not a cause-and-effect relationship. Parasitic worms cause both these under-eye bags and insomnia, as well as other sleep-related diseases such as snoring (due to pharyngeal inflammation and swelling) and sleep apnea. Trying to sleep more without treating the underlying infection will not eliminate the bags, and the insomnia will remain. Treating the parasites improves both conditions. This is relevant because a person's health status and thus his/her extent of lung disease are correlated roughly to the size of his/her under-eye bags. The extent of your disease, a disease that we all have, is written plainly on your face, for all the world to see, if the world knows what to look for. Everyone eventually dies of parasitic worm disease, unless they die of something else first, even if lung disease is not the primary problem. Everything that you think of as old age is actually caused by parasitic worms—parasite-free old age has never existed. I now breathe and sleep better than I have for 35 years, and my eye bags are nearly shriveled up.

In regard to sleep apnea, these worms and the body's immune response to them will be shown to be the major factor in the obesity epidemic (along with high-fructose corn syrup, which worms apparently love, and which should be strongly restricted in foods and especially beverages), thus absolving obese people of much (not always all) of the blame for their condition. (Interestingly, sodas that use sugar instead of high-fructose corn syrup [e.g. Jarritos brand] appear not to be especially unhealthy.) It breaks my heart to see people on television saying that they had a stroke, etc., and are dying because of their smoking habit when their doctors are obviously (to me) wrong, falsely blaming the patients for their misfortunes, when it is obvious to me that parasitic worms are the true cause of their problems. These poor, dying people suffer enough without receiving such calumny from people who are misguided, ignorant, and wrong. (Sorry! No offense intended. It's just objective, constructive criticism.) A lot of time and money have been spent to fight smoking, and have you noticed that as smoking rates decline, obesity increases, and life expectancy has dropped proportionately? It's not all due to smoking decline alone, of course, but you probably never made that association, before, right? I contend that parasitic worms are the reason. I have never smoked and probably will never smoke, and I certainly don't recommend smoking as therapy (at this point), but facts are facts, whether we like them or not. I strongly urged my mother to quit smoking, and when she did, she died shortly thereafter, and I believe that my delusional desire to help her quit smoking accelerated that process. I feel very bad about that, but I, like you, was misinformed by well-intentioned but wrong medical authorities. I was young and ignorant; I am no longer either. Talk about irony! It turns out that perhaps the tobacco industry was right.

In summation, the time has come to recognize these lung and other diseases for what they really are, and not some incurable autoimmune/ genetic disease of questionable existence. (Although I do not question the existence of genetic diseases in general.) While analyses of autoimmune diseases are filled with unexplained contradictions, using parasitic infection such as the hyperinfection syndrome of strongyloidiasis as an explanation for these many diseases adds a great deal of clarity to an understanding of medicine—many puzzles are answered and solved, and previously idiopathic diseases are cured. Remember Occam's Razor. Treating these many diseases appropriately as parasitic worm infections will dramatically lower health care costs and improve the quality of care as well as the quality of life for billions of people around the world. The time has come for a paradigm shift. Let's get started.

REFERENCES

1   World Cancer Report. Lyon, France: International Agency for Research on Cancer, 2008.
2   Mathers CD, Loncar D (November 2006). "Projections of Global Mortality and Burden of Disease from 2002 to 2030". *PLoS Med.* **3** (11): e442. doi:10.1371/ journal.pmed.0030442. PMC 1664601. PMID 17132052.
3   Rabe KF, Hurd S, Anzueto A et al. (2007). "Global Strategy for the Diagnosis, Management, and Prevention of Chronic Obstructive Pulmonary Disease: GOLD Executive Summary". *Am. J. Respir. Crit. Care Med.* **176** (6): 532–55. doi:10.1164/rccm.200703-456SO. PMID 17507545.
4   MedicineNet.com — COPD causes, accessed 6/11/13.

5   Young RP, Hopkins RJ, Christmas T, Black PN, Metcalf P, Gamble GD (August 2009). "COPD prevalence is increased in lung cancer, independent of age, sex and smoking history". *Eur. Respir. J.* **34** (2): 380–6. doi:10.1183/09031936.00144208. PMID 19196816.

6   Lamprecht B, McBurnie MA, Vollmer WM. Chest. 2011 April; 139(4): 752–763. Published online 2011 March 26. doi:10.1378/chest.10-1253.

7   Suckling, B; Johnson, MM; Chin, R; *Nutrition, Respiratory Function, and Disease;* In: Shils M, Shike M, Ross AC, et al, eds. *Modern Nutrition in Health and Disease, 10th edition.* Lippincott Williams & Wilkins, Philadelphia, 2005. 1470, 1467.

8   http://www.lung.org/stop-smoking/about-smoking/health-effects/smoking. html. American Lung Association. Accessed 6/11/13.

9   http://en.wikipedia.org/wiki/Immunoglobulin_E. Accessed 6/11/13.

10   Tierney LM, McPhee SJ, Papadakis MA. *Strongyloidiasis.* In: *Current Medical Diagnosis and Treatment 2002, 41st edition.* Lange Medical Books/McGraw-Hill, New York, 2002. 1526-1528.

11   Ash LR, Orihel TC. *Intestinal Helminths.* In: Balows A, Hausler WJ, Herrmann KL, et al, eds. *Manual of Clinical Microbiology, Fifth Edition.* American Society for Microbiology, Washington, D.C. 786-788.

12   Ash LR, Orihel TC. *Intestinal Helminths.* In: Balows A, Hausler WJ, Herrmann KL, et al, eds. *Manual of Clinical Microbiology, Fifth Edition.* American Society for Microbiology, Washington, D.C. 787.

13   http://emedicine.medscape.com/article/229312-overview, Accessed 6/11/13.

14   http://iceh.uws.edu.au/fact_sheets/FS_strongyloidiasis.html.   Accessed 6/11/13.

15   Ash LR, Orihel TC. *Intestinal Helminths.* In: Balows A, Hausler WJ, Herrmann KL, et al, eds. *Manual of Clinical Microbiology, Fifth Edition.* American Society for Microbiology, Washington, D.C. 787.

16   Turnlund JR. *Copper.* In: Shils M, Shike M, Ross AC, et al, eds. *Modern Nutrition in Health and Disease, 10th edition.* Lippincott Williams & Wilkins, Philadelphia, 2005. 287ff.

17   Turnlund JR. *Copper.* In: Shils M, Shike M, Ross AC, et al, eds. *Modern Nutrition in Health and Disease, 10th edition.* Lippincott Williams & Wilkins, Philadelphia, 2005. 289.

18   http://en.wikipedia.org/wiki/Coffee. Accessed March 21, 2013.

19   http://en.wikipedia.org/wiki/Tobacco_smoking. Accessed March 21, 2013.

20   http://en.wikipedia.org/wiki/Nicotine. Accessed March 21, 2013.

21   http://en.wikipedia.org/wiki/Caffeine. Accessed March 21, 2013.

22   http://en.wikipedia.org/wiki/Acetaldehyde. Accessed 6/11/13.

23   Dr. Orien Tulp, of the University of Science, Arts, and Technology, from a lecture.

24   Eckhert CD. *Other Trace Elements.* In: Shils M, Shike M, Ross AC, et al, eds. *Modern Nutrition in Health and Disease, 10th edition.* Lippincott Williams & Wilkins, Philadelphia, 2005. 339-341.

25 http://en.wikipedia.org/wiki/Immunoglobulin_E. Accessed 6/11/13.

26 Cannon JG. *Cytokines and Eicosanoids*. In: Shils M, Shike M, Ross AC, et al, eds. *Modern Nutrition in Health and Disease, 10th edition*. Lippincott Williams & Wilkins, Philadelphia, 2005. 663.

27 Brodsky IG. *Hormones and Growth Factors*. In: Shils M, Shike M, Ross AC, et al, eds. *Modern Nutrition in Health and Disease, 10th edition*. Lippincott Williams & Wilkins, Philadelphia, 2005. 640.

28 Brodsky IG. *Hormones and Growth Factors*. In: Shils M, Shike M, Ross AC, et al, eds. *Modern Nutrition in Health and Disease, 10th edition*. Lippincott Williams & Wilkins, Philadelphia, 2005. 649.

29 Cannon JG. *Cytokines and Eicosanoids*. In: Shils M, Shike M, Ross AC, et al, eds. *Modern Nutrition in Health and Disease, 10th edition*. Lippincott Williams & Wilkins, Philadelphia, 2005. 663.

30 Cannon JG. *Cytokines and Eicosanoids*. In: Shils M, Shike M, Ross AC, et al, eds. *Modern Nutrition in Health and Disease, 10th edition*. Lippincott Williams & Wilkins, Philadelphia, 2005. 657.

31 http://en.wikipedia.org/wiki/Nicotine. Accessed March 21, 2013.

32 http://www.winespectator.com/webfeature/show/id/Wine-Drinkers-More-Likely-to-Live-Longer-Study-Finds_3532. Accessed 6/11/13.

33 Tierney LM, McPhee SJ, Papadakis MA. *Enterobiasis*. In: *Current Medical Diagnosis and Treatment 2002, 41st edition*. Lange Medical Books/McGraw-Hill, New York, 2002. 1518-1520.

34 Tierney LM, McPhee SJ, Papadakis MA. *Strongyloidiasis*. In: *Current Medical Diagnosis and Treatment 2002, 41st edition*. Lange Medical Books/McGraw-Hill, New York, 2002. 1526-1528.

35 Tierney LM, McPhee SJ, Papadakis MA. *Strongyloidiasis*. In: *Current Medical Diagnosis and Treatment 2002, 41st edition*. Lange Medical Books/McGraw-Hill, New York, 2002. 1527.

36 Orihel TC, Ash LR. *Tissue Helminths*. In: Balows A, Hausler WJ, Herrmann KL, et al, eds. *Manual of Clinical Microbiology, Fifth Edition*. American Society for Microbiology, Washington, D.C. 779.

Chapter 9

Curing Chronic Disease 2.0

A more clinical argument against the focus on patient safety is that medical injuries do not cause as many deaths as errors of omission ... Our health care system fails with embarrassing frequency to provide medical interventions known to benefit patients. — Thomas H Lee M.D. (2002)

Parasitic worm infection is the cause of a great many modem illnesses that have defied explanation until now—neurological, autoimmune, psychiatric, and various other diseases including so-called "normal aging." The most-likely parasite is *Strongyloides stercoralis,* an endemic roundworm first discovered in 1876; it is less than 3 mm long.[1-3] A partial list of worm caused diseases that I, directly or through others, have cured using only anti-worm therapy includes: Parkinson's disease, autism (Asperger's), dementia (probably Alzheimer's), multiple sclerosis, gluten allergy/celiac disease, miscellaneous allergies, chronic heartburn (GERD), Chronic Fatigue Syndrome, osteoporosis, depression, anxiety, type-2 diabetes mellitus, acne, and many others. There are about 80 autoimmune diseases, and I now believe that parasitic worm infection might cause them all. I believe that everybody has these worms and that they have been present in humans from the very beginning of human history, being transmitted in utero— worms nicely explain women's decreasing fertility with age and a variety of gynecological problems. Also, everyone seems to have "floaters" in their eyes that are actually parasitic worms, and "normal" aging looks identical to parasitic worm proliferation.

I believe that these organisms, causing parasitic encephalitis (brain inflammation/infection by parasites), will be shown to cause a great many psychological/psychiatric/neurological diseases, running the gamut from amyotrophic lateral sclerosis (Lou Gehrig's disease) and anxiety/depression to epilepsy, schizophrenia, and Tourette's. When I look at James Holmes, the schizophrenic Colorado theater killer (and/or John Hinckley Jr., who shot President Reagan, or Jared Loughner [Gaby Giffords'shooter], or Adam Lanza [Sandy Hook]), I see a man/victim with parasitic encephalitis, who probably could have been prevented from his crime if he had been properly treated with anti-worm medicine. I see that Alzheimer's disease and autism are essentially the same disease, just occurring at different life stages. My friend cured his grandson's autism within 24 hours (at least in a behavioral sense) using anti-worm therapy alone (mebendazole), which exceeded even my expectations.

Most people unconsciously use a variety of anthelminthic (anti-worm) substances every day. Smoking has been found to improve Alzheimer's, Parkinson's, schizophrenia, emphysema, obesity, ulcerative colitis, and anxiety, and the reason is because nicotine is a naturally-occurring pesticide![4] Smokers are subconsciously treating their worm infections! This is why smokers deteriorate/gain weight when they quit smoking. Therefore, the war on smoking is misguided and wrong, and based on ignorance (although I do hate second-hand smoke). Also, many modern, idiopathic (meaning "due to unknown causes") inflammatory diseases are treated with immunosuppression; this is wrong, very wrong, because the diseases (COPD, asthma, back pain, dry eyes, etc.) are caused by worms, and suppressing the immune system makes parasitic diseases worse in the long run despite achieving short-term relief (e.g. cortisone shots, cyclosporine eye drops); idiopathic inflammatory diseases should be treated with anthelminthic drugs and lifestyles first. Worms even explain why people shrink as they age—worms are eating away at your spine, causing disability, pain, and eventually death via both direct destruction and competition for nutrients. Another popular, natural anti-parasitic drug is caffeine,[5] found naturally in coffee plants and various tea plants, used by approximately 70% of the population worldwide. Also, the reason that marijuana treats glaucoma is because it too has anti-parasitic (anti-acetylcholinesterase) properties.[6] Finally, alcohol also appears to be toxic to worms, although none of these treatments is curative.

I have cured type-2 diabetes mellitus using anthelminthic drugs without dieting, weight loss, exercise, or surgery, and I have no doubt at all that type-1 (adult-onset/insulin dependent) diabetes mellitus has the same cause and treatment. Interestingly, the parasite, being transmitted before birth, has its first point of contact with its victim inside the liver (via the umbilical vein from the placenta), whereas infections acquired after birth mostly have to pass through either the skin or the throat—more-difficult propositions. Consequently, these worms have easy access to their new host and, after multiplying in the glucose-and-nutrition-rich liver, can easily slide down the hepatic duct to the duodenum (upper intestine) and then chew their way directly into the immediately-adjacent pancreas, where they, multiplying very slowly by parthenogenesis[3,7] (non-sexual reproduction), eventually cause type-1 diabetes mellitus by both destroying pancreatic tissue (insulin-producing beta cells) and by causing a copper deficiency, copper being essential for insulin (and hormone and nerve) function;[8,9] from the duodenum, they can easily cause Crohn's disease or ulcerative colitis. Conversely, the worms can instead enter the bloodstream from the liver, travel up to the heart and lungs, and then go anywhere, including the brain. These worms can penetrate any tissue,[3] including bones, which presumably leads to a variety of blood diseases such as leukemia, myelodysplastic syndrome, and multiple myeloma, but they seem to like the lymphatic system especially well, which makes sense, because when fatty food enters the body, it first passes into chylomicrons in the lymphatic system. Worms cause massive, grotesque lymph node enlargement that is often mistaken for fat, and quite possibly lymphoma, I suspect. (They would be killed by full-body radiation during a bone marrow transplant—probably explaining this mechanism of therapy in cancer treatment.) Because the worms multiply very slowly, these infections require decades to reach critical mass—10-20 years for type-1 diabetes; 20 or so years for schizophrenia; 30 or so years for type-2 diabetes; 50 or so years for emphysema; 50-60+ years for Parkinson's and Alzheimer's diseases, but much depends on presumably random movement of the parasites, dietary habits, inheritance, and other factors.

Interestingly, lymphoma, diabetes mellitus, and autoimmune diseases, et al. are all associated with celiac disease/gluten allergy;[10,11] therefore, I believe that lymphoma, leukemias and other blood disorders will certainly be linked to these worms and the resultant mineral deficiencies, as well as being a factor in other cancers. These worms are the best explanation for a congenital, inherited disease with a non-genetic mode of transmission. For instance, the incidence of schizophrenia in a second identical twin when the first twin is so afflicted is 40 percent,[12] while twin inheritance of allergies ranges from 40% to 70%[11,13] —worm infection as a cause of both diseases explains this concordance nicely. (Worms increase immunoglobulin E [IgE], and thus allergies, e.g. gluten allergy, hayfever.) People often say that diabetes is a modem disease, but it has been known for approximately 2,000 years,[14] so that assertion is wrong—for example, I believe that I can show that my family has had parasitic worms for more than 500 years, having a long history of diabetes, glaucoma, Alzheimer's, Parkinson's, personality disorders, etc. These infections cause mineral deficiencies, of which copper and magnesium may be the most important. Copper deficiency is the reason that dopamine fails to be made in Parkinson's disease, thereby blocking the first of the two chemical reactions (tyrosinase enzyme needs copper to make DOPA from tyrosine, an amino acid) in the dopamine synthesis pathway. (FYI: I am the first person to cure Parkinson's disease, et al. Also: Serotonin synthesis follows the same exact pathway, except starting with the amino acid tryptophan rather than tyrosine.) In fact, many of the signs and symptoms of aging are synonymous with copper deficiency: skin wrinkles (lysyl oxidase impairment), pale skin and gray hair (melanin impairment), hormone deficiencies (aldosterone, cortisol, and sex hormone synthesis blockage; blockage of catecholamine synthesis—tyrosinase and dopamine beta-hydroxylase, etc.), blockage of vitamin D synthesis and subsequent increase in cancer (blockage of hydroxylation and oxidation of cholecalciferol to 1,25 dihydroxycholecalciferol [vitamin D]), vision loss (retinal melanin impairment in macular degeneration) and of course many other reactions.[15-17]

In an informal survey of people who wear glasses, every person who wears glasses seems to have floaters in his/her vision- i.e. floating spots that move around in the visual field (best observed by looking at a clear blue sky). Pretty much everyone of a certain age also has these floaters. Despite whatever ridiculous explanation your doctor has given you, floaters are worms. See how some of them are squiggle-shaped? And see how some of them look like masses of worms? Worms are now and have always been part of "normal" aging, causing a great many diseases, and not just in Africa or Thailand. I believe that the under-eye bags, wrinkles, and/or dark circles that are ubiquitous after a certain age are pathognomonic ("uniquely identifying") for parasitic worm infection, having achieved amelioration of these with anti-worm therapy. Up until the 1930s, the average human lifespan was only approximately 35 years; because strongyloides infections have been documented to last for 65 years,[18] they rarely caused problems until relatively recently, when people suddenly started living longer, due largely to the recognition of bacteria beginning around 1878, the subsequent introduction of refrigeration and improved food handling and transportation (to reduce bacterial growth), the introduction of antiseptic/aseptic surgery (by Joseph Lister) starting around 1865 or so, and the successful treatments of bacterial infections starting around 1939 with antibacterial drugs such as Prontosil (Gerhard Domagk, Nobel Prize, 1939). Now, parasitic worms are probably the leading cause of death among the elderly, causing a wide variety of "age-related" diseases.

Parasitic worms are the most important health problem in western societies. This discovery will have enormous implications not only for public health, but also for family violence, social services, criminal justice systems, and probably even politics worldwide. I believe that treating schizophrenics for worms will prevent the mass murders that are increasingly common. Perhaps recognizing the significance of these mind-altering parasitic worms as a cause of anger and conflict between groups of people will lead to a new era of world peace. Go to RobertSFarmerMD.com for more information— Robert S. Farmer, M.D.

Summary: Unrecognized parasitic worm infections are the major cause of disease in western civilization. They cause a variety of neurological, psychological, autoimmune, allergic, and other diseases that have until now eluded definitive cure by the medical establishment, including psychiatric/neurological diseases that cause psychosis and lead to mass murder. Most of these diseases are aging-related, but some, such as autism and allergies, manifest at young ages. Treating these diseases is not difficult or dangerous and largely involves the use of standard anti-worm medicines and lifestyle and dietary modifications. Initial improvement may occur rapidly, but definitive treatment requires years of therapy.

REFERENCES

1 Farmer, RS. Parkinson's Explained. http://www.RobertSFarmerMD.com. accessed Nov. 17, 2012.
2 http://en.wikipedia.org/wiki/Strongyloides, accessed Dec.16, 2012.
3 Ash IR, Orihel TC. Intestinal Helminths. In: Balows A, Hausler WJ, Herrmann KI, Isenberg HD, Shadomy HJ, eds. Manual of Clinical Microbiology, Fifth Edition. American Society for Microbiology, Washington, D.C. 786-788.
4 http://en.wikipedia.org/wiki/Nicotine, accessed Dec. 17, 2012.
5 http://en.wikipedia.org/wiki/Caffeine, accessed Dec. 17, 2012.
6 http://en.wikipedia.org/wiki/MedicaLmarijuana, accessed Dec. 16, 2012.
7 Tierney IM, McPhee SJ, Papadakis MA. Current Medical Diagnosis and Treatment 2002, 41st edition. Lange Medical Books/McGraw-Hill, New York, 2002. 1526-1528.
8 Fields M. Nutritional Factors Adversely Influencing the Glucose/Insulin System; Journal of the American College of Nutrition, August, 1998, http://www.jacn.org/content/17/4/317.full, accessed Nov. 17,2012.
9 Fields M, Reiser S, Smith JC Jr. Effect of copper or insulin in diabetic copper-deficient rats. US National library of Medicine, National Institutes of Health. http://www.ncbLnlm.nih.gov/pubmed/6344092, accessed Nov 17, 2012.
10 Connon JJ. Celiac Disease. In: Shils M, Shike M, Ross AC, et al, eds. Modern Nutrition in Health and Disease, 10th edition. Lippincott Williams & Wilkins, Philadelphia, 2005. 1222.
11 Taylor SL, Hefle SL. Food Allergies and Intolerances. In: Shils M, Shike M, Ross AC, et al, eds. Modern Nutrition in Health and Disease, 10th edition. Lippincott Williams & Wilkins, Philadelphia, 2005. 1521.
12 http://en.wikipedia.org/wiki/Schizophrenia, accessed Nov. 16, 2012.
13 http://en.wikipedia.org/wiki/Allergy, accessed Dec. 16, 2012.
14 Anderson JW. Diabetes Mellitus: Medical Nutrition Therapy. In: Shils M, Shike M, Ross AC, et al, eds. Modern Nutrition in Health and Disease, 10th edition. Lippincott Williams & Wilkins, Philadelphia, 2005. 1053.
15 Turnlund JR. Copper. In: Shils M, Shike M, Ross AC, et al, eds. Modern Nutrition in Health and Disease, 10th edition. Lippincott Williams & Wilkins, Philadelphia, 2005. 286ff.
16 Champe PC, Harvey RA. Biochemistry, 2nd edition. J.B. Lippincott Co., Philadelphia, 1994. 167, etc.

17   USDA National Agricultural library website, Copper: http://fnic.nal.usda.gov/
    dietaryguidance/dri-reports/vitamin-vitamin-k-arsenic-boron-chromium-
    copper-iodine-ironmanganese; http://www.nal.usda.gov/fnic/DRI/ /DRI_
    Vitamin_A/224-257 lSO.pdf; accessed Nov. 17, 2012.

18   http://en.wikipedia.org/wiki/Strongyloides, accessed Dec.16, 2012.

19   Siddiqui AA, Berk Sl. Diagnosis of Strongyloides stercoralis Infection. Oxford
    Journals. http://cid.oxfordjournals.org/content/33/7/1040.full.pdf, downloaded
    Nov 19, 2012.

Chapter 10

The Cure for Diabetes Mellitus

The emergence of endocrinology allowed the development in the 1920s of insulin treatments which saved the lives of diabetics. But one must not assume that diabetes then went away: no cure has been found for that still poorly understood disease, and it continues to spread as a consequence of western lifestyles. Indeed one could argue that the problem is now worse than when insulin treatment was discovered. — Roy Porter, *The Greatest Benefit to Mankind*

Diabetes Mellitus (DM) affects approximately 20 million people in the U.S. alone and is projected to affect 300 million people worldwide by 2025.[1] It is a problem that potentially affects every person on earth.

The usual explanation for DM is that obesity causes type-2 diabetes via insulin resistance, and some unknown virus or gene causes type-1 diabetes by destroying the pancreatic beta cells that make insulin. I will demonstrate that these beliefs are false or at the very least overly simplistic, and I will explain the real mechanisms and how I have cured type-2 diabetes and will certainly cure type-1 diabetes as well. I will also demonstrate that genetics most likely has absolutely no role in DM whatsoever.

In 1928, Dr. Elliot P. Joslin, a famous diabetes expert and author, proclaimed, "with an excess of fat diabetes begins, and from an excess of fat diabetics die."[2] I used to think that this was just wrong, but I now see that it's merely overly simplistic—the absolute, most-superficial level of analysis—kindergarten medicine. Therefore, it is very unfortunate that everyone has believed this simplistic, primitive, and basically erroneous proclamation for the last 85 years, placing the blame on diabetics inappropriately. Because most people believe that fat (this term will be used as a synonym for cholesterol and triglycerides) causes type-2 DM, let's begin our inquiry by examining this allegation.

First, does fat cause diabetes? The best answer is no, it does not *cause* DM, however it is an integral intermediary in the DM process; nevertheless, finding the root cause of DM requires a deeper level of analysis. It turns out that fat is not simply a storage form of energy; it is also quite active metabolically, secreting a variety of cytokines (chemical messengers), including tumor necrosis factor-alpha (TNF-α) and interleukin-1 beta (IL-1β).[3] These inflammatory cytokines are part of the immune system's response to infection, and in the case of DM ... severe, chronic infection.[4] TNF-α., among other functions, is responsible for creating insulin resistance, its most usual function for this being in pregnant women, to ensure that the fetus receives enough glucose.[5] IL-1β, on the other hand, is responsible for decreasing insulin secretion in order to raise blood sugar.[6] Why would the body do this? The most probable reason is that prokaryotic bacteria can be killed osmotically by high glucose levels, which then attract water and cause lysis (rupture) of the infective cell. Hyperglycemia (high blood glucose) appears to be part of the body's normal immune response system.[7] This mechanism is most likely used mainly after other killing mechanisms such as antibodies and phagocytic (killing) cells have failed to eradicate the infection. Unfortunately, it is probably not effective with eukaryotic infections. Now, then, we have determined that glucose-raising cytokines secreted by fat cells are part of the immune system's response to severe infection, and that hyperglycemia is a predictable and normal part of that immune response. So then, let's move to a deeper level of analysis and ask, "What causes the formation of fat cells?"

Typically, overweight people are accused of causing their obesity by overeating, eating unhealthy foods, and just generally having low willpower, such as by avoiding exercise. All of this is wrong (usually). It turns out that fat, in the form of triglycerides (trigs) and very low density lipoproteins (VLDLs), are also created as part of the body's response to severe infection (e.g. foam cells [macrophages] in atherosclerosis),[8] which makes enormous sense when one considers that the secretions from fat cells are also part of the immune response. I have personally observed people who should not have been fat, based on their caloric intake, and yet they were severely obese—it didn't make a damned bit of sense! But if we consider that the fat is not mainly a storage form of energy, but an essential response to chronic, severe infection, the equation changes completely, making perfect sense! However, we are then left with the question, "What infection could cause this?"

Most people look at diabetics without seeing any sign of infection. At most, they might see signs of "old age" —bags under the eyes; 'floaters' in the eyes; chronic pain; pale, wrinkled, loose skin; spinal degeneration; gray hair; obesity; hormone deficiencies. While these may or may not be signs of old age, they are definitely signs of chronic, severe infection with parasitic worms such as *Strongyloides stercoralis* and the copper deficiencies that these chronic infections cause![9] I have cured type-2 DM by treating for these parasites, and the evidence shows that everyone on earth has this unrecognized infection, acquired from our barefooted ancestors in the distant past and passed to us in the womb by our unknowing and powerless mothers.

Interestingly, an infection acquired in the womb can travel from the placenta down the umbilical vein and be implanted directly into the fetal liver without having to pass through the skin in the usual sense. Once implanted in the liver, the tiny (less than 3 mm long[10]) worms have full access to a nutritious feast, where they can dine and frolic merrily, until they need to reproduce (which they do very slowly, often by asexual reproduction[11]) and move away from the parents. Where might they go? Well, the liver is the Grand Central Station of the body, so they might go down the hepatic duct toward the duodenum (small intestine), but perhaps stop halfway to tour the pancreas, where they can happily cause type-1 DM over many years by destroying beta cells. Alternatively, the liver directly contacts the gastroesophageal junction, so they can cause gastroesophageal reflux disease (GERD), stomach problems, esophagitis, bronchitis, COPD, asthma, etc., or they can go to the right kidney, the colon, the small intestine—all of which the liver contacts directly. This, then, explains ulcerative colitis (i.e. inflammatory bowel disease—which is improved by smoking[12] because tobacco has anti-parasitic chemicals such as nicotine and acetaldehyde[13]), irritable bowel syndrome, gluten allergy (celiac disease), nephritic/nephrotic kidney disease, and probably Crohn's, etc. (I've cured most of these "incurable" diseases by treating for parasites.) The tiny worms also seem to love the lymphatic system, causing extreme and grotesque enlargement of lymph nodes, thus I believe that they are a major factor in lymphoma, leukemia, and other forms of cancer, especially breast cancer, and they are definitely responsible for severe, debilitating back pain by destroying the collagen in the spine and back muscles. Through both direct and indirect effects, they are responsible for perhaps almost every chronic disease, including Parkinson's disease, Multiple Sclerosis, Alzheimer's, and various kinds of mental illness such as schizophrenia, which, like most parasitic diseases, usually has adult onset (schizophrenia,[14] Parkinson's,[15] and Alzheimer's disease[16] are also improved by smoking,[14-16] although doctors usually confuse cause and effect). Medical students are taught that there are two causes of increased Immunoglobulin E (IgE) in the body: 1) parasites/worms, and 2) allergies/autoimmune diseases. The problem with this is that parasites/worms *cause* allergies and so-called "autoimmune" diseases, reducing the causes of increased IgE to one: immune response to parasites. I've cured allergies by reducing parasitic worms. (Just try it!) Additionally, IgE stimulates mast cells to release TNF-α,[17] thereby establishing a direct cause-and-effect link to hyperglycemia/DM from parasitic worm infection. To reiterate: parasitic worms infect the body before birth, causing secretion of fats (trigs and VLDLs) and IgE by the immune system; the fat cells and mast cells are then stimulated by the worm-induced IgE to secrete TNF-α and presumably Il-1β. These cause hyperglycemia/DM in an ineffective attempt to eradicate the eukaryotic infection after other immune system mechanisms have failed. In some cases, the worms also move from their starting point in the liver into the nearby pancreas and destroy beta cells directly. The patient then usually dies from the combination of the unrecognized infection and possibly the hyperglycemia as a complication. This is also complicated by mineral deficiencies, especially in regard to copper, which is essential for both the immune system and insulin function.[18,19] I've seen this many, many times. It is so obvious in hindsight.

Therefore, obesity does not in fact cause DM; fat/obesity is merely an intermediate effect in the process of unrecognized, severe, chronic parasitic infection that eventually leads to DM and death, so understanding and curing DM requires a deeper level of analysis than the superficial treatments of diet, weight loss, and insulin. Doctors have until now confused cause and effect in their analysis of this disease, to the detriment of their many patients over these last 137 years since *Strongyloides stercoralis* was first recognized in 1876 by Dr. Louis Normand.[20] Genetics and viruses apparently have no involvement in this disease. Anti-parasitic drugs such as mebendazole, ivermectin, and praziquantel are all useful in curing DM, but in much longer than recommended doses, for months or years, with frequent rest periods, due to Jarisch-Herxheimer reactions. There has been a great deal of needless suffering that could have been mitigated if only we had recognized this sooner. Also, curing diabetes mellitus will cure diabetic retinopathy, one of the leading causes of blindness. For more information, please go to :www. RobertSFarmerMD.com.

All truth passes through three stages: First, it is ridiculed. Second, it is violently opposed. Third, it is accepted as being self-evident. —Arthur Schoepenhauer

Summary: Diabetes Mellitus is caused by a well-defined immunological response to unrecognized, severe, chronic parasitic worm infection. This infection is complicated by severe mineral deficiencies, especially in regard to copper. Eliminating or reducing the parasites and replacing the depleted minerals cure diabetes mellitus.

REFERENCES

1   Anderson JW. *Diabetes Mellitus: Medical Nutrition Therapy.* In: Shils M, Shike M, Ross AC, et al, eds. *Modern Nutrition in Health and Disease, 10th edition.* Lippincott Williams & Wilkins, Philadelphia, 2005. 1043.

2   Anderson JW. *Diabetes Mellitus: Medical Nutrition Therapy.* In: Shils M, Shike M, Ross AC, et al,eds. *Modern Nutrition in Health and Disease, 10th edition.* Lippincott Williams & Wilkins, Philadelphia, 2005. Page 1053.

3   Brodsky IG. *Hormones and Growth Factors.* In: Shils M, Shike M, Ross AC, et al, eds. *Modern Nutrition in Health and Disease, 10th edition.* Lippincott Williams & Wilkins, Philadelphia, 2005. 640.

4   Cannon JG. *Cytokines and Eicosanoids.* In: Shils M, Shike M, Ross AC, et al, eds. *Modern Nutrition in Health and Disease, 10th edition.* Lippincott Williams & Wilkins, Philadelphia, 2005. 663.

5   Brodsky IG. *Hormones and Growth Factors.* In: Shils M, Shike M, Ross AC, et al, eds. *Modern Nutrition in Health and Disease, 10th edition.* Lippincott Williams & Wilkins, Philadelphia, 2005. 649.

6   Cannon JG. *Cytokines and Eicosanoids.* In: Shils M, Shike M, Ross AC, et al, eds. *Modern Nutrition in Health and Disease, 10th edition.* Lippincott Williams & Wilkins, Philadelphia, 2005. 663.

7   Cannon JG. *Cytokines and Eicosanoids.* In: Shils M, Shike M, Ross AC, et al, eds. *Modern Nutrition in Health and Disease, 10th edition.* Lippincott Williams & Wilkins, Philadelphia, 2005. 663.

8   Cannon JG. *Cytokines and Eicosanoids.* In: Shils M, Shike M, Ross AC, et al, eds. *Modern Nutrition in Health and Disease, 10th edition.* Lippincott Williams & Wilkins, Philadelphia, 2005. 663.

9   Turnlund JR. *Copper.* In: Shils M, Shike M, Ross AC, et al, eds. *Modern Nutrition in Health and Disease, 10th edition.* Lippincott Williams & Wilkins, Philadelphia, 2005. 287-288.

10   Ash LR, Orihel TC. *Intestinal Helminths.* In: Balows A, Hausler WJ, Herrmann KL, Isenberg HD, Shadomy HJ, eds. *Manual of Clinical Microbiology, Fifth Edition.* American Society for Microbiology, Washington, D.C. 786.

11   Tierney LM, McPhee SJ, Papadakis MA. *Current Medical Diagnosis and Treatment 2002, 41st edition.* Lange Medical Books/McGraw-Hill, New York, 2002. 1526-1528.

12   Griffiths AM. *Inflammatory Bowel Disease.* In: Shils M, Shike M, Ross AC, et al, eds. *Modern Nutrition in Health and Disease, 10th edition.* Lippincott Williams & Wilkins, Philadelphia, 2005. 1210.

13   *Acetaldehyde.* http://en.wikipedia.org/wiki/Acetaldehyde. Accessed Mar. 7, 2013.

14   *Schizophrenia.* http://en.wikipedia.org/wiki/Schizophrenia Accessed Mar. 7, 2013.

15   de Lau LML, Breteler MMB. *The Lancet Neurology,* Volume 5, Issue 6, *Epidemiology of Parkinson's disease.* 525-535. (Used via Wikipedia, accessed Mar. 7, 2013.)

16   *Alzheimer's Disease,* http://en.wikipedia.org/wiki/ Alzheimer%27s_disease. Accessed Mar. 7, 2013.

17   Cannon JG. *Cytokines and Eicosanoids.* In: Shils M, Shike M, Ross AC, et al, eds. *Modern Nutrition in Health and Disease, 10th edition.* Lippincott Williams & Wilkins, Philadelphia, 2005. 657.

18   Fields M. *Nutritional Factors Adversely Influencing the Glucose/Insulin System;* Journal of the American College of Nutrition, August, 1998, http://www.jacn.org/content/17/4/317.full, accessed Nov. 17, 2012.

19   Fields M, Reiser S, Smith JC Jr. *Effect of copper or insulin in diabetic copper-deficient rats.* US National Library of Medicine, National Institutes of Health. http://www.ncbi.nlm.nih.gov/pubmed/6344092, accessed Nov 17, 2012.

20   Strongyloides; http://en.wikipedia.org/wiki/Strongyloides, accessed Dec. 16, 2012.

Chapter 11

Curing Blindness

All truth passes through three stages. First, it is ridiculed. Second, it is violently opposed. Third, it is accepted as being self-evident. —Arthur Schopenhauer

10/23/12 - Updated 11/15/12

Introduction

The leading causes of blindness are glaucoma, diabetic retinopathy, and macular degeneration, according to *Current Medical Diagnosis and Treatment*. This cited source implies that this information applies to the U.S.A., as the article goes on to say that the leading causes of blindness worldwide are cataracts, trachoma, leprosy, onchocerciasis, and xerophthalmia (dry eyes). Glaucoma, diabetic retinopathy, and macular degeneration can all be explained nicely by parasitic worm infection, most probably by *strongyloides stercoralis* or something very similar. I believe that this infection will most likely also explain cataracts and xerophthalmia, as well. River Blindness (onchocerciasis) and trachoma need no real explanation, as the pathologies (parasitic worms and Chlamydia, respectively) and treatments (ivermectin and doxycycline, respectively) are well known. This also applies to Leprosy, although perhaps the treatments for this disease are not completely effective against the known pathogen *Mycobacterium leprae*, which causes leprosy, and better treatments for this known infection may still be necessary.

Before launching into an examination of specific diseases that cause blindness, it will be necessary to examine what I believe is the most overlooked disease on earth: strongyloides parasitic worm infection, or strongyloidiasis. Based on my personal experience and based on my observations of other people, I believe that virtually everyone has congenital strongyloides infection, or an infection that is very similar. For me personally, I have had more than 40 diseases including Parkinson's disease, Multiple Sclerosis, and Amyotrophic Lateral Sclerosis, among many others, and I have cured these supposedly incurable diseases by using anti-worm (anthelminthic) treatments. These three diseases in particular are, of course, neurological diseases, and furthermore, I am certain that they are primarily-neurological infections, or else I would not have been able to treat them successfully with my limited resources, using the methods that I did in fact use. In addition, it is important to realize that this infection, in addition to causing direct destruction of neurological tissues, including peripheral nerves as well as optical nerves and destruction of the brain and its optical centers as well, also causes extreme mineral deficiencies, especially in regard to copper, which is essential for formation of the myelin sheath of nerves, as well as being essential for the synthesis of essential hormones and various other substances, and being essential for the structural integrity of every structure in the body because of its essentiality for the enzyme lysyl oxidase, which cross-links collagen, which forms virtually every tissue in the body. In fact, a deficiency of copper appears to be almost synonymous with aging, as it causes gray hair, wrinkles of the skin, osteoporosis, collagen degeneration (e.g. ligaments), hormone depletion (including: vitamin D, which is important or essential for fighting cancer; and insulin, a deficiency of which causes diabetes mellitus), and retinal dysfunction, cataracts, macular degeneration, and other causes of blindness. Specifically, in relation to vision, copper is essential for the synthesis of melanin that allows the retina to perceive light. Without copper, there is no melanin, and without melanin, there is no functionality of the retina, and without the retina, there is no vision. Now, then, if we operate on the provisional assumption that virtually everyone is infected with strongyloides in the womb, explaining many diseases becomes quite simple. Let us examine the first three major causes of blindness: glaucoma, diabetic retinopathy, and macular degeneration.

Glaucoma

Glaucoma is caused by a buildup of pressure in the eye. The mechanism of action is believed to be due to blockage of Schlemm's canal and other vessels that drain aqueous fluid from the eye. The tissue surrounding the ducts swells for unknown reasons, compressing the ducts and preventing drainage of fluid. This buildup of fluid causes increased pressure that eventually compresses the nerves of the eye and can lead to death of the optic nerve, and, thus, blindness.

The swelling of the tissue that compresses Schlemm's canal and other vessels that drain the eye is easily explained by infection with strongyloides. It is well-known that swelling of the face, scalp, and throat occurs in neurological diseases such as Parkinson's disease and ALS, and I have personally experienced this. Killing the infective parasites reduces the swelling and allows normal function to return to the affected tissues. In the case of glaucoma, aqueous drainage via Schlemm's canal etc. would resume.

Perhaps you're thinking that this sounds ridiculous. If parasitic worms were the cause of glaucoma, then how would that infection be treated? Well, one way to treat it would be to use an insecticide/pesticide. What would be suitable? An acetylcholinesterase inhibitor is a common pesticide. And what is historically and commonly used to treat glaucoma? An acetylcholinesterase inhibitor! Doctors have unwittingly been using pesticides, basically, to treat glaucoma for decades, yet without realizing the mechanism of action! True, most pesticides are irreversible inhibitors of acetylcholinesterase, while physostigmine, a usual treatment for glaucoma, is a reversible inhibitor of acetylcholinesterase, but the principle is the same—the goal is to paralyze or kill the infectious organism. The problem, then, is that the treatment, physostigmine (or pilocarpine) is reversible, meaning that the patient must keep taking the drug to keep the parasitic infection low enough to prevent significant eye damage. A better treatment would be to use praziquantel, an anti-worm drug, dissolved in ethyl alcohol (drinkable alcohol) to ensure penetration of the central nervous system, which is what I successfully used on myself. Additionally, copper supplements will help to heal and to return function to the nervous tissue, as well as being a substance that the body uses naturally to fight parasites. I take about 10 mg/day of copper, more or less, which is much more than the inadequate Recommended Daily Allowance, and which some authorities regard as the upper safe limit; although I often take more than 10 mg/day, it probably averages out to around ten when one includes the days when I forget to take my supplements. The toxic level of copper has been shown to be in excess of 30 mg/day for more than three years, in one observational study. Therefore, ten milligrams per day seems to be well within a safe range. In any event, copper is easily excreted by the liver. Additionally, the daily level of copper intake recommended by the World Health Organization (1.3 mg/day) is approximately twice the level of the Recommended Daily Allowance in the U.S., which is a mere 0.7-0.9 mg/ day (*Copper in Health, Wikipedia; Modern Nutrition in Health and Disease*). Upon even superficial examination, it should be clear that Americans are generally deficient in copper, a deficiency that in any event is worsened by unrecognized, chronic parasitic infection. On top of this, high-fructose corn syrup is a relatively recent addition to the American diet, and studies have found that high-fructose corn syrup binds copper, preventing it from being absorbed, and flushing it out of the body, further depleting Americans of copper. Also, vitamin C is a very popular wintertime preventive treatment for illness, but taking excessive levels of vitamin C also binds and flushes copper out of the body. Vitamin C should never be taken without copper for this reason, but virtually no one is aware of this fact. Cow's milk is also a dietary substance that is so low in copper that it appears to absorb copper from the body and cause it to be lost; although milk is nutritious in general, it is not good for copper levels in the body, in excess.

Are there any anti-parasitic, medicinal substances that laypeople commonly use to treat glaucoma? Yes, tobacco inhibits parasitic worms (nicotine is used as a pesticide), but far more people prefer to treat their glaucoma with marijuana, an herb that has known anti-parasitic properties, and which is now quite popular and a substantial part of the government tax base in California and Colorado. Other popular anti-parasitic medicinal substances include coffee, which is the highest source of boron, an anti-parasitic mineral, in the American diet (6.7% of daily boron *intake—Modern Nutrition in Health and Disease*), as well as being the highest source of caffeine, another anti-parasitic substance (*Caffeine, Wikipedia*). Interestingly, people who drink coffee live 10-16% longer than people who do not drink coffee, despite the finding that coffee has more than 1,000 chemicals in it, many of which are alleged to be unhealthy *(Coffee, Wikipedia)*. Apparently, those chemicals *are* very unhealthy—for parasites!

Other factors that support parasitic infection as a cause of glaucoma are:

- Glaucoma, like parasitic worms, is associated with headaches.
- Migraine headaches are often exacerbated by caffeine; caffeine kills parasites such as worms. Therefore the caffeine-induced headache pain is most-likely caused by toxins released from the dying parasitic worms—a phenomenon known as a Jarisch-Herxheimer reaction.
- Glaucoma is sometimes attributed to perforation of Bruch's membrane, which could easily be caused by parasitic worms.
- Parasitic worms such as *Onchocerca volvulus* cause many cases of blindness in Africa.
- One common cause of Closed-Angle Glaucoma is contraction of a membrane *(Merck Manual)*. This could easily be caused by parasitic worms, which have caused sclerosis.
- Another common cause of Closed-Angle Glaucoma is central retinal vein occlusion *(Merck Manual)*. This could easily be caused by parasitic worms.
- Many of the treatments for glaucoma (beta-blockers, laser surgeries, etc) treat the signs and symptoms rather than the root cause of glaucoma. Physostigmine is an exception to this phenomenon, yet it is nevertheless insufficiently adequate for cure, in part because it is not an irreversible inhibitor of acetylcholinesterase.
- Isofluorophate, an irreversible acetylcholinesterase inhibitor, is a treatment for glaucoma and probably functions as a pesticide.
- Glaucoma, like parasitic worm infections, tends to affect people who are middle-aged and elderly more than the young, although in neither case are the young completely immune. Parasitic strongyloides populations usually require decades to reach critical mass.
- Many people treat glaucoma with marijuana, a known inhibitor of parasitic worms. These treatments have become virtually mainstream in places such as California and Colorado, yet no one

seems to recognize the mechanism of action—parasitic worm inhibition!

- Glaucoma, like parasitic worms and diabetes mellitus, runs in my family, indicating a probable non-genetic inheritance pattern that is consistent with congenital parasitic worm infection.

Unfortunately, many of the treatments that have recently been used to treat glaucoma, such as beta-blockers, carbonic anhydrase inhibitors, prostaglandin analogs, and laser surgery, do nothing to treat the underlying cause of glaucoma, allowing the underlying, causative parasitic infection to progress without inhibition, eventually leading to blindness.

I think that the likelihood of glaucoma being caused by parasitic worm infection is overwhelming, especially when one considers that I reversed my own, minimal visual decline via anthelminthic therapy, and that anthelminthic therapy causes a therapeutic-feeling inflammation and clearing of the eyes, with shrinkage of the eyeballs that is detectable upon palpation, indicating a probable decrease in intraophthalmic pressure, along with a subsequent improvement in vision.

Treating glaucoma with alcohol-praziquantel in combination with physostigmine or perhaps isofluorophate, an irreversible acetylcholinesterase inhibitor that has been used to treat glaucoma, would most-likely produce the best results by quite possibly creating improvement synergistically. Perhaps an intravenous preparation of alcohol-praziquantel could eliminate the parasites more quickly, although an increased Jarisch-Herxheimer reaction would be a matter of concern.

Diabetic Retinopathy

Millions and millions of dollars have been spent allegedly to find a cure for diabetes mellitus. I have a friend who has Type 2 diabetes mellitus, and he has reported a dramatic improvement since beginning anthelminthic therapy, but lots of people report reversal of their Type 2 diabetes mellitus upon weight loss, although my friend didn't have to lose weight to experience his improvement. His improved insulin resistance, lowered blood sugar, and improved general sense of well-being was caused solely by praziquantel, an anthelminthic drug, dissolved in potable (ethyl) alcohol. Weight loss generally follows anthelminthic therapy, though—a factor that will be shown to be critical in fighting the American obesity epidemic. He was kind of a bad patient, because he kept forgetting to take his copper supplements, thereby demonstrating that substantial improvement is possible using anthelminthic medication alone.

It should go without saying that in order to treat diabetic retinopathy, one must first treat diabetes mellitus. So then, let us cure diabetes mellitus, right here and right now.

Aside from the dramatic improvement in blood glucose control that killing presumed strongyloides infection provides (eliminating competition for nutrients), it is worth noting that parasitic worms can easily have an even more direct role in the onset of Type 1 (insulin injection-dependent) diabetes mellitus. An interesting fact about strongyloides parasitic worms is that they can penetrate any tissue in the body, even including bones. One of the favorite places for these worms is the gastrointestinal tract, which is where these infections seem to begin, which makes sense if they are inherited in the womb. The umbilical vein that feeds the fetus leads to the liver; the ducts that drain the liver enter the duodenum; the duodenum contacts the pancreas—where insulin is made—directly. It does not take a great deal of imagination, then, to see that the pesky little parasites can easily chew their way from the duodenum or the hepatic duct directly into the immediately adjacent pancreas, where they can happily destroy a variety of cells, including alpha and beta cells, which are essential for maintaining blood glucose in the healthy range, thus quite feasibly leading to juvenile-onset diabetes mellitus (type 1). Of course, beta cells are the ones that produce insulin, and because we know that strongyloides or other parasitic infections cause decreased levels of copper (which is the mechanism for Parkinson's disease, copper being required for tyrosinase to form DOPA from tyrosine), copper being essential for the formation of insulin, it should be obvious to us that parasitic worms can cause Type 1 diabetes mellitus by direct destruction of beta cells as well as by depleting copper in the pancreas and throughout the body more generally and thereby also inhibiting formation of insulin by undamaged beta cells. Parasitic worm infection would explain the heretofore unexplained finding that many diabetics have increased levels of copper in the damaged areas of their pancreases—the body uses copper to kill parasites! Researchers have hypothesized that, because elevated levels of copper are found near damaged pancreatic cells in diabetics, that perhaps excessive levels of copper cause diabetes! Once again, they have confused cause with effect! Copper does not cause diabetes; copper release is the body's response to parasitic infection! Diabetes mellitus is caused by parasitic worms! Doctors' misguided efforts to cure diabetes and cancer by reducing copper levels will only make those diseases worse, by inhibiting hormone production and the immune system, of which copper is a critical component. Copper is also essential for insulin function.

Undoubtedly, depletion of other minerals is also involved in causing diabetes, as is protein depletion by competition for nutrients by the voracious little scavengers, because insulin is composed of numerous amino acids linked together. In particular, magnesium is essential for the formation of adenosine triphosphate (ATP), which is an essential form of energy in the body. Whereas many "experts" speculate that a virus causes loss of beta cell function, what this usually means in practice is that nobody knows what causes the decline, so the default response is that the problem must be either genetics or an unknown virus. Actually, the evidence for parasitic worm infection is much more compelling.

The usual treatment for diabetes mellitus is to replace the insulin that is not being made. Similarly, the treatment for Parkinson's disease is to replace the DOPA/dopamine that is not being made. Both are bandage therapies that fail to address the root causes. However, I have cured Parkinson's disease by killing the worms with alcohol-praziquantel and replacing the copper to eliminate the deficiency that prevents the formation of DOPA from tyrosine (see my article *Parkinson's Explained* on my website). Likewise, then, curing diabetes mellitus should be possible by killing the worms with alcohol-praziquantel or some other anthelminthic drug and replacing the copper and other minerals that are depleted in the pancreas. Supplementation with protein supplements such as soy protein powder should also be helpful, as the parasites seem to have a voracious appetite for proteins. Curing diabetes mellitus then should lead to reversal of diabetic retinopathy. Undoubtedly, the parasites are also in the eyes (I've seen worms in my own eyes) and the brain, and definitive, curative therapy will take years, and has never been accomplished in the history of the world, although, fortunately, complete eradication of the parasites, while desirable, is not strictly necessary, as the most odious manifestations of the infection will diminish or disappear long before the infection is completely eradicated.

There seems to be a common misconception that nerve damage cannot be healed. This is wrong, as peripheral nerves can heal at the rate of one millimeter per week, and contrary to earlier beliefs, the brain itself can also be remodeled. Therefore, I have little doubt that diabetic retinopathy can be reversed in cases of brief duration if the root causes are adequately treated, because the human body has an amazing capacity to heal itself when given the proper assistance. I am also hopeful that long-standing cases of blindness due to nerve damage might also be reversed and cured, although that can certainly not be guaranteed. However, once the infectious organisms are killed or dramatically reduced, there is always the possibility of stem cell treatment to return vision to the patient.

As an aside, I used to experience severe, debilitating hypoglycemia on a several-times-a-day basis, requiring me to carry sweet snacks or drinks with me constantly, indicating a probable degeneration or inhibition of my glucagon-secreting alpha cells, as well as inhibition of my counter-regulatory catecholamine hormone system, which requires copper. In addition, I had severe disseminated inflammation and salt craving/deficiency and weakness indicating deficiencies of cortisol and aldosterone, two cholesterol-based hormones that require copper for the hydroxylation and oxidation reactions that form them. Now, subsequent to anthelminthic therapy, I never experience these hypoglycemic attacks until I've fasted for 24 hours or so. Also, my inflammation and salt cravings are dramatically decreased or eliminated. These findings are solely due to anthelminthic therapy with mineral supplementation. It's a cure that no other doctor on earth has accomplished.

Macular Degeneration

My father has macular degeneration, and he is loaded with parasitic worms. Using Occam's Razor, the probability that his macular degeneration is not caused by parasites seems ridiculously small, especially when he has many additional signs and symptoms that point toward helminthic parasitic infection, such as spinal degeneration and general back pain, gastrointestinal problems, Asperger's syndrome, two personality disorders, prostate problems, early-stage Alzheimer's dementia, and much more.

Parasitic worms such as strongyloides are known to infect the eyes, and my eyes in particular have benefitted greatly from anti-worm therapy. I have seen floaters in my vision that clearly looked like worms, having a long, cylindrical, worm-like shape with three 180-degree bends and then tapering out of my range of focus at each end. I am certain that the copper deficiency that these worms cause will also be shown to be critical in the etiology of macular degeneration.

Copper is essential to form the melanin that allows the retina to function. This necessity is clearly demonstrated in albinism, a genetically-caused disease in which patients have numerous, severe visual problems due to their inability to use copper properly (defect in tyrosinase, also known as tyrosine hydroxylase, the enzyme that is affected by copper deficiency in Parkinson's disease, which I have cured by treating for worms). Dopamine is a catecholamine hormone/neurotransmitter that is essential for proper retinal function, and many of the rods in the retina are activated by G-proteins which are often stimulated by epinephrine (adrenalin), which is a catecholamine hormone that is made from dopamine, and which therefore also requires copper for its formation, copper and vitamin C apparently being necessary for many or all hydroxylation reactions in the human body. Dopamine is also required for the function of a lesser-known optical pigment known as melanopsin:

Dopamine (DA) is a factor in the regulation of melanopsin mRNA in ipRGCs. Because DA synthesis and release in the rat retina are under the control of rods and cones, it appears that rods and cones, in conjunction with or possibly with the exclusion of direct circadian or photic input, control transcription of melanopsin.

*ipRGC: intrinsically photosensitive Retinal Ganglion Cells.
A Sakamoto K, Liu C, Kasamatsu M, Pozdeyev NV, Iuvone PM, Tosini G. Dopamine regulates melanopsin mRNA expression in intrinsically photosensitive retinal ganglion cells. Eur J Neurosci. 2005; 22: 3129-3136. (Quoted from *Melanopsin*, *Wikipedia*.)

I also feel certain that the drusen that are characteristically found in macular degeneration will be found to be related to parasitic worm infection; these drusen or bodies similar to drusen have been found in patients with Alzheimer's dementia, which is another disease that is certainly caused by parasitic worm infection. Drusen are also related to atherosclerosis, which is also undoubtedly related to parasitic worm infection, as atherosclerosis has been shown to be reduced by ingesting turmeric with black pepper—a well-known anti-parasitic worm treatment; the black pepper contains a chemical named piperine that increases the anthelminthic properties of turmeric (active ingredient: curcumin, see *Curcumin, Wikipedia; Turmeric, Wikipedia*) synergistically many times. I personally have had some kind of undefined dementia (probably Alzheimer's) that was reversed in the course of my anti-worm treatment, and I have no doubt that these diseases are related.

Another interesting finding is that zeaxanthin, a pigment in the eye, is made from beta-carotene by hydroxylating beta-carotene; hydroxylation reactions apparently always involve copper and vitamin C, although this may not be absolutely definitive. It is also not clear whether the human body makes zeaxanthin from beta carotene or whether zeaxanthin must be supplied in the diet. However, if zeaxanthin can be made in the human body, there is a very good chance that it is made with the assistance of copper. If we additionally consider that zeaxanthin is an isomer of lutein, another retinal pigment, we might speculate whether some interconversion occurs. If, as is probable, zeaxanthin can not be formed by humans, it can be formed from beta-carotene in the human body by some of the microorganisms that humans harbor, such as *Escherichia coli* and the yeast *Saccharomyces cerevisiae* (*Zeaxanthin, Wikipedia*). In this case, adequate copper and vitamin C in the human body would still undoubtedly be required for normal visual function. This will undoubtedly be important in treating night blindness.

It has been noted that macular degeneration has an association with smoking, leading to the supposition that smoking causes macular degeneration. This is backwards! Macular degeneration (and emphysema/COPD) causes smoking! Once again, doctors have confused cause with effect. Here's what happens: parasitic worms start out in the maternal/fetal bloodstream (umbilical vein) and move into the fetal liver, then the duodenum and the intestines, and they slowly move up to the head and eyes, passing through the lungs and heart on the way. People with macular degeneration smoke because smoking is one of the few ways that people have to inhibit parasitic worms, which they feel especially when their lungs are heavily infected! Smoking and other forms of tobacco are known to inhibit worms (along with coffee), due at least in part to nicotine, a natural pesticide. People don't know that they are treating their parasitic worm infections, but they unconsciously adopt behaviors that make themselves feel better, and reducing their parasitic worm levels makes them feel better! Have you ever wondered why smoking improves Parkinson's disease, schizophrenia, emphysema, and anxiety, among other diseases? It is because they are caused by parasitic worms, and parasitic worms are inhibited by smoking! Other dietary items that inhibit worms are coffee, cinnamon, cloves, garlic, olive oil, turmeric, black pepper, salt, and numerous other spices, also some fruits such as pineapple, cherries, berries, and vegetables such as carrots and celery, etc. People think that the reason that the Mediterranean diet treats bad cholesterol and/or atherosclerosis is because of some mysterious direct effect on cholesterol, but I believe that it is because of the known fact that garlic and olive oil really kills or inhibits worms with surprisingly high effectiveness. Try some spaghetti loaded up with one or two big cloves of garlic, pour on olive oil (no tomato sauce), add some parmesan cheese, and see what I mean! There is a Jarisch-Herxheimer reaction that is very powerful (which you might interpret as the pain of indigestion). Add turmeric, cinnamon, and black pepper for an extra-strong effect. Wash it down with some coffee. Also, a 200 mg caffeine pill usually makes me very sleepy from the Jarisch-Herxheimer effect caused by the dying parasites; obviously this is not what most people would anticipate from an alleged stimulant drug. (Although 400 mg at once just feels scary.) However, caffeine is primarily a natural pesticide that plants make for their own protection (*Caffeine, Wikipedia*) before it is a stimulant used by people. It has a strong effect against parasitic worms, which explains why coffee use is endemic in countries all around the world, including the U.S.A. This also explains why nicotine is second only to caffeine in popularity as a stimulant—they both inhibit or kill worms.

Other supporting data suggesting that parasites cause macular degeneration, uveitis, and various eye diseases:

- Wine is believed to have some protective effect against macular degeneration. It also inhibits worms, as does alcohol more generally. This explains the so-called French paradox.
- Parasitic infection and competition for nutrients in the eye could lead to neovascularization and hypervascularization, which are often associated with macular degeneration.
- Parasitic eye infection can lead to dry, gritty-feeling eyes (xerostomia), which are often associated with macular degeneration.
- Parasites cause allergies, including hayfever and gluten allergy, and raise levels of immunoglobulin E. These affect the eyes, too. This can lead to xerostomia.
- Photophobia is often believed to be caused by viruses. I believe that accusing viruses of causation is code for "we don't know." I've proved definitively that photophobia can be caused by parasitic worms, because I've cured the extreme photosensitivity that I had for about 40 years. I doubt that that has ever been done before.
- Floaters are thought to have no significance, but I believe that I have shown that they do. They are worms.
- I used to experience flashing lights upon closing my eyes at night, and now I don't. The only explanation is that I have is parasitic worm infection that has been dramatically improved by praziquantel etc.
- The steroid/glucocorticoid treatments (also, cyclosporine/Restasis, etc.) that many people use to treat inflammation of their eyes (and back, e.g. chronic back pain) will make their eye disease (and back pain and general health) worse in the long term by allowing the parasitic worms to proliferate. Some eye doctor somewhere must have made this observation, I would imagine, but maybe not. I believe that my father's treatment of his eye disease with blephamide for probably a decade or so has led to his personality disorders (two of them) and/or Alzheimer's dementia, and general decline, all due to parasitic worm infection (he, like his grandson, also has Asperger's syndrome, a form of autism, which, using Occam's razor, is undoubtedly caused by parasitic worms; see my article *Autism*

on my website). Similarly, when my mother finally quit smoking, she went rapidly downhill and died, which is not what the health authorities of the day told me/the public would happen! And I feel terrible and generally duped about that! But I know better, now.

- My Aunt Beth had what I believe to be a semi-classic worm infection: she was very skinny for many years, more than average; at an advanced age, she had surgery for cataracts; she went into a nursing home, became obese and demented, and died at 90 years of age. Of course, she lived longer than most, but I believe that her cataracts were probably related to her worm infection and copper depletion. The obesity was caused by a proliferation of parasites that eventually overwhelmed her body, depleted her copper, and killed her. I think that the evidence, although some of it is circumstantial, is nevertheless overwhelming for an etiology of parasitic worm disease, quite possibly *Stronglyoides stercoralis*, as a cause of blindness.

Cataracts

As I suggested above, I believe that worms probably cause cataracts. My evidence, however, is slim. Cataracts, of course, have numerous causes, such as Chlamydia bacteria (e.g. in trachoma) and herpes/varicella viruses, but I certainly think that adding parasitic worms to the differential diagnosis is a safe bet.

A More Personal Set of Observations

Pretty much everyone's eyes get worse as they age, and some people's eyes aren't so good to begin with, apparently. My eyes were basically fine, or so I thought, until my late forties. However, upon closer investigation, I had had excessive, even extreme, light sensitivity since at least the eighth grade, and probably before that. In eighth or ninth grade, I dropped out of Algebra because the classroom gave me a headache every day that I was in it, which was every other day, and the occurrence was quite consistent. The classroom faced south, into the sunlight, and the walls were painted white, and the ceiling was white, and the floor was basically white, and there was a ridiculous number of 4-ft-long fluorescent light bulbs lighting the room, and it was the fluorescent light bulbs that were the main problem, although, in all honesty, studying algebra, and in a room with no girls, was probably less-than-helpful in this regard, too. It is well-known that many people get headaches from fluorescent lights because of the way that they flicker—I seem to recall that they flicker about 60 times per second. The teacher refused to turn off the lights, even though the classroom was flooded with sunlight; because this was in Anchorage, Alaska, the sun was relatively low in the sky, thus the sunlight entered the room through the windows farther than it does at lower latitudes, creating better illumination, making the use of additional lighting unnecessary—and aggravating my headaches. In any event, except for my light sensitivity, which seemed to be in my brain rather than just in my eyes; my eyes seemed fine.

This light sensitivity continued to get worse, sometimes causing me to wear my very dark sunglasses and hats indoors, for instance in my college classrooms, which I'm sure made people think that I had been smoking marijuana. However, I've never smoked marijuana, despite numerous opportunities. (Although, marijuana has anti-parasitic properties.) Then, in my thirties or so, I noticed "floaters" in my vision, which are somewhat dark spots that appear in a variety of places in one's visual field; also, they tend to move around, appearing to float, hence the name "floaters." I went to the Wilmer Eye Clinic at Johns Hopkins Hospital, and they told me that no one knew what caused them. They may have offered some ridiculous explanation about how there are gaps between cells or something in the vitreous humor that cause it, but the main points that I took away were that no one knew what caused them and that virtually everyone has them. I did not find that reassuring.

Next, I noticed that my vision was getting blurry in my late forties, and it was clearly related to some allergies of recent onset. My vision would fluctuate depending on how active my allergies were on any given day. Clearly, any blurriness was not due to nerve degeneration or cataracts, because those don't fluctuate—they get steadily worse. Around this time, I also noticed that I had acquired very large, dark circles under my eyes. The skin over those circles also looked a bit loose, although they weren't really bags, such as what some people have, although bags were definitely not far away from forming. My eyeballs would often feel irritated, and I would often massage them with my fingers. Sometimes they felt dry and sandy, sort of, although I did not have dry eyes consistently. Sometimes at night, I would see flashing lights after I closed my eyes, especially after taking voriconazole, an anti-fungal drug. Flashing lights after closing one's eyes are accepted as a sign of possible retinal detachment, which is a very serious condition. Also, when I lay back outside on a lawn and looked at the clear blue sky, I could see my floaters better, and I noticed that some of my floaters looked like worms, a long, cylindrical worm-like structure with three 180-degree bends in them, but no clear beginning or end—they looked like the ends just got blurry and faded away. Some of them looked like they might be tangles of worms, but there was no good way to get a good look at the more complicated floaters—they would move away whenever I looked at them, moving constantly.

Of course, as you might be aware from reading the other articles on my website, I was having numerous other health problems, too. Eventually, after years of antibacterial and anti-fungal therapies, I started treating myself presumptively for parasitic worm infection, after closely studying the article on strongyloidiasis in *Current Medical Diagnosis and Treatment,* et al. That was the first key! As all of my other health problems started to diminish, my eye problems also diminished. Now, after treating for my case of presumed *strongyloides stercoralis* infection, my vision is probably *20/10* again, which is what it usually was, although it still fluctuates a little bit. A rather exciting finding is that for the first time in more than 25 or 30+ years, I can go outside without sunglasses and/or a hat—which feels very strange to me, now. On a related note, the first few anti-worm treatments that I used did not seem to be clearly effective, although there may have been some mild effects. Two of the less effective drugs that I used were ivermectin and albendazole. Years earlier, I had been able to relieve some gastrointestinal problems with mebendazole and metronidazole in combination or sequence, but the prohibitively expensive cost of mebendazole ($35/pill) and the numerous doses that would be required for adequate treatment prevented me from trying it more recently. I'm sure that the only reason that mebendazole is so expensive is because of price-gouging by the manufacturers and/or sellers, as I have heard from a traveler to Africa that mebendazole costs ten cents per pill there. There is a strong possibility that the reason that these other drugs were not effective for me is because I did not dissolve them in alcohol, as I did with praziquantel; therefore, those drugs would not be expected to penetrate various tissues including the blood-brain barrier as effectively as the alcohol-praziquantel preparation. This would turn out to be the second key—combining the anti-worm drug with alcohol to allow penetration into the central nervous system is critical! Consequently, the usefulness of those first drugs for treating these various diseases should not be excluded summarily; they may well be quite useful, and, additionally, the probability that different kinds of parasites with different chemical sensitivities are involved cannot be excluded, either—but the drugs need to penetrate the central nervous system. Alcohol, of course, would also aid the drug in penetrating the pancreas, presumably, inasmuch as pancreatitis is apparently/allegedly often caused by excessive use of alcohol; this finding would be useful in the treatment of diabetes mellitus and therefore diabetic retinopathy.

As I use my anti-worm protocol, I've noticed increased immune system activity in my eyes—increased pus and tears—as my vision improves. I have fewer problems that are attributable to allergies, which I have virtually none of, anymore, having eliminated my hayfever and my extremely debilitating gluten allergy. (How many doctors besides me know how to cure allergies?) The dark circles under my eyes have diminished dramatically, and I expect that they will disappear before too long. As the circles clear up, my acne has also cleared up, so that my complexion is better than it has been in about 40 years—currently with no active acne, after a brief period of increased acne as the parasites initially died off. Now, when I massage my eyeballs, they feel smaller than they were, indicating reduced intraocular pressure, which is the main concern in glaucoma. The eyes, or at least the retina, are considered to be a direct extension of the brain, and as my eyes have cleared up, I've been having fewer brain problems, although, really, the brain problems mostly cleared up before my eyes did. Still, everything is getting better, slowly.

These improvements have no explanation other than anti-worm therapy and mineral replacement, especially with copper.

My father has had many of the same health and eye problems that I have had, and I believe that I inherited my worm infection from my parents. He has been diagnosed with macular degeneration (see my article *Macular Degeneration*), which I believe was made worse by years of inappropriate therapy with blephamide, an anti-inflammatory, immune system-inhibiting glucocorticoid (steroid) ophthalmic drug. I found that using blephamide made my eyes feel better, too, although fortunately I tried it for only a short time. That is the problem with these drugs: by inhibiting the immune system, they reduce inflammation and make the eyes feel better in the short term; however, by allowing the worms to escape the immune system, the worms can proliferate, which makes the patient worse in the long term, so the patient continues to deteriorate, which can lead to blindness, mental illness, and dementia, among other diseases, and not until decades later, when the past medical history is long lost, buried, and/or forgotten. The treatment that makes my eyes both feel better and work better is anti-parasitic worm therapy, specifically praziquantel in an alcohol solution with copper (separately) and soy protein powder supplementation, in addition to multi-vitamins and multi-minerals. Many people will at first be allergic to whey and/or soy (I was allergic to whey for a year or two), but the allergies will remit upon continuing anthelminthic therapy. (Soy protein powder is better than whey.)

I believe that dark circles under the eyes may be pathognomonic for parasitic worm infection, meaning that the cause of them can be nothing else. Look around you and see how many people have dark circles under their eyes, then tell me whether you think that parasitic worm infection is common. Everyone over a certain age has them! The famous Dr. Oz, who is slightly younger than I am, has very bad bags under his eyes, despite the fact that he is one of the most health-conscious people anywhere! I've noticed that some of Dr. Oz's favorite supplements are green coffee bean extract, garlic, and turmeric. He doesn't know it, but the reason that these make him feel better is because they are treating his worm infection!

In addition, I believe that parasitic worms often cause many more cases of blindness than they are given credit for. It is well-known that *onchocerca volvulus* causes many cases of blindness in Africa (e.g. River Blindness), but cultural arrogance in European and North American cultures prevents us from recognizing the significance of worms in western civilizations. I've noticed that the swelling of my pharynx that caused me to snore very loudly (probable obstructive sleep apnea) as well as the misery-inducing nasal stuffiness that plagued me at bedtime for many years have been essentially eliminated now that my parasitic worm infection is on a dramatic decline, and I have no doubt at all that parasitic worm infection of the eyes would cause glaucoma by inducing swelling, thereby inhibiting drainage of aqueous fluid due to blockage of Schlemm's canal and the other drainage systems of the eye. Because worms love the eyes and because the eyes and the pharynx are very close together, the association seems obvious. In fact, glaucoma runs in my family, which I believe I have already established has had parasitic worm infections for generations, and probably from the beginning of human history. My great aunt Mary on my father's side had glaucoma, and I was told that she was not the first family member to have it. My aunt Louise, actually my father's cousin, had type 1 diabetes mellitus, which I was also told ran in our family. My paternal grandfather suffered from dementia that is now believed to be Alzheimer's disease, another disease that is undoubtedly of helminthic origin. My alleged ancestor King Henry VIII of England (of the house of Tudor) had numerous psychological problems, was very obese, and is alleged by modern investigators to have had type 2 diabetes mellitus *(King Henry VIII Wikipedia)—all* probably caused by worms. Finally, of course, I have a bad case of worms, and my nephew (my father's grandson) has Asperger's syndrome, a type of autism that I feel certain is caused by parasitic worms. All of this—four modern generations plus one generation from history—taken in combination, suggests that congenital parasitic worm infection has been in my family for more than 500 years, and undoubtedly from the beginning of the human race and its early ancestors.

Additionally, parasitic worms can directly attack nerves such as the retinal nerve and the optic nerve, and they can no doubt travel along these nerves into the brain and damage visual centers of the brain directly (probably rarely, though, because of their posterior location), because the eyes are considered to be an extension of the brain. Parasitic worm infection can explain a great many cases of blindness, the causes of which have previously gone unrecognized. Also, the treatment for glaucoma is physostigmine, which is a reversible acetylcholinesterase inhibitor. Using this drug, basically a pesticide, in your eyes would probably kill, paralyze, or otherwise inhibit the worms that cause glaucoma, but because physostigmine is reversible, one could predict that the effects on the worms would wear off, which is why people need to take physostigmine continuously, rather than just once. A more effective treatment would probably be to use isofluorophate, an irreversible acetylcholinesterase inhibitor, in combination with alcohol-praziquantel. It all makes perfect sense. I don't think that anyone really knows exactly why cholinergic agonists (e.g. acetylcholinesterase inhibitors) such as physostigmine work, but its effect as a pesticide that kills parasitic worms is the best explanation. Unfortunately, many of the newer drugs that are being used to treat glaucoma probably do not inhibit worms, a situation which will therefore allow the eye diseases to progress, despite the apparent temporary resolution of signs and symptoms.

Fortunately, these parasitic worms, *Strongyloides stercoralis*, reproduce by parthenogenesis, which is a relatively slow means of reproduction (unlike, say, *Ancylostoma duodenale*, the females of which reproduce by laying 30,000 eggs in a day *[Wikipedia]*), which is why they have eluded diagnosis for so long. This is why, for the entire history of human beings, parasitic worm infection has been mistaken for old age. The worms, which can allegedly live for 70 years or more, reach a critical mass as humans achieve advanced ages, thus causing a mistaken diagnosis of "old age." *Strongyloides stercoralis* infections have been documented to last for 65 years, while until 80 or so years ago, the average human lifespan was only about 35 years.

Strongyloides infections are life-long unless treated. Therefore, I believe that no one has ever really died of old age, because parasitic worms undoubtedly killed them first, unless something else killed them before the parasites did. This applies to dogs and other mammals, as well. I've seen dogs die from obvious worm infection that veterinarians were unable to recognize, which became clear to me in retrospect. For all of human existence, human lifespan has been limited by parasitic disease; this continues today.

Miscellaneous Observations

Today (10/23/12) on *The Doctors* TV show, Dr. Stork pointed out that children with obstructive sleep apnea have something like a 40 percent chance of having behavioral problems. The allegation is that oxygen deprivation causes Attention Deficit Hyperactivity Disorder, etc. This is wrong. Obstructive sleep apnea and the adenoid and tonsil swelling that is associated with it is caused by parasitic worms; they also cause psychiatric problems such as ADHD, personality disorders, and even schizophrenia and dementia. I don't know how so many people can get this so wrong! Read a medical book, sometime, Dr. Stork!

Also, Dr. Stork mentioned that approximately 30 percent of people have an allergy to red wine, so people are now inventing non-allergenic red wine. Are you kidding me?!! As I have already mentioned, red wine antagonizes worms, apparently in part because of sulfites that are in wine. This allergy is caused by parasitic worm infection. Once you know what to look for, you will see parasitic worm infection everywhere, and it is very aggravating!

I recently (11/2/12—this is an update) saw a report on ABC News that researchers have established a link between parents stress levels and their children's obesity, a link that they blamed on fast food. Wrong! It's an interesting finding, but the link is parasitic worms! I recently heard on TV that people who don't smoke or who quit smoking live 10-15 years longer than people who don't smoke, according to researchers. This is wrong. For one thing, we all know that these researchers have not followed people around for 50 or 100 years until they died in order to make these observations. They are making statistical assumptions to use in their calculations in order to arrive at their conclusions. Their main problem, then, is that they have an unrecognized confounding variable. The thing that these researchers have failed to take into consideration is that people who smoke have parasitic worm infections that are worse than are the parasitic worm infections of people who don't smoke. People who need to smoke to control their infections who then quit will die sooner than they would have if they continued to smoke—like my mother, they will lose years from their lives, not add to them. Of course, people who either have light parasitic worm infections or who somehow control their infections with diet rather than smoking will live longer, but that is not a fair or direct evaluation of the effects of smoking. It is impossible to do a truly accurate comparison of smoking effects versus not smoking or quitting smoking, because you cannot test for two or three different outcomes on the same person; in other words, you cannot wait for old Bill to die in order to test one set of lifestyle choices, then resurrect him postmortem to restart the clock from the same age to test another set of lifestyle choices using the same exact starting point to see if you can achieve a different outcome—it's just impossible to eliminate confounding variables, so statisticians make the best estimates that they can. Unfortunately, this time, they screwed up, most likely because 1) like everyone but me and one doctor friend, they are ignorant of parasitic worm infections, and 2) there is a lot of irrational, wrong prejudice against smoking. I remember hearing on the TV news about a guy who lived to be 103 years old, "despite" the fact that he smoked heavily and drank coffee all day long; he didn't live to be 103 despite his bad habits, he lived to be 103 *because* of them. His "bad" habits kill worms.

I've always hated smoking, and for many years, I erroneously blamed it for my mother's health problems. However, explaining why people smoke has been a question that has always bugged me. The answer has always been that people become addicted, but that still doesn't explain why they initially started, why they continued until they got addicted, and in any event why they don't try harder to quit. I've always just felt that people who smoke must be stupid; even if they were otherwise brilliant, they had a stupid, self-destructive streak that no one had been able to explain. This was a problem for me, because I really didn't want to believe that my mother was stupid—in several ways, her intelligence was clearly above-average—but there just seemed to be no other explanation; "everyone" "knew" (double emphasis) that smoking was bad for people. How could anyone explain why an otherwise intelligent person would do something as dumb as smoking, literally poisoning oneself 400 or so times a day with every puff? This dilemma continued to bug me for decades. I just had to keep looking for a logical explanation. Eventually, I decided that literally everyone—8 billion people—was (were) wrong. Somewhere and somehow, I decided that because smoking improves Parkinson's disease, emphysema, schizophrenia (I worked in a psych ward once), anxiety, and worms, the answer had to be parasitic infection, which proved to be true in every case that I could examine. I always have a picture in my mind of a moose with one of those parasitic brain diseases (Moose Brain Worm, possibly *Paraphostrongylus tenuis*) that causes them to run blindly through the woods, running directly over anything and anyone in their way until they die—this is the same expression that James Holmes, the schizophrenic Colorado theater killer, had on his face in court. It all seems so obvious (self-evident) in retrospect.

It was always said of Albert Einstein that he was always genuinely happy when someone pointed out one of his mistakes, because from that point forward, he knew better than before. Similarly, I believe that one of the best defining criteria for intelligence is the rapidity with which a person changes his/her mind when confronted with superior evidence. Using this, my own criterion, I passed this test reasonably well, as I had to completely reverse the opinions and prejudices that I had had about smoking and smokers for numerous decades in order to arrive at these conclusions, and they were astonishing, at first. I felt really quite dumb for not noticing it sooner. Still, though, I assuage my embarrassment by reminding myself that 8 billion other people missed it completely.

I don't want to defend smoking, particularly, but the truth is the truth. Smoking, while not exactly healthy, is less unhealthy than not smoking, for people with serious parasitic infections that are not controlled by other means. And there is the rub—there are better and therefore healthier ways of treating parasitic worm infection, although due to the costs of seeing a doctor and paying grossly inflated prescription drug prices, the question arises whether medical treatment is actually cheaper than smoking. This probably involves subjective calculations about the value of one's health. However, at this point, the question is moot, because there are currently only two doctors on earth who recognize the true cause of these many diseases. I certainly hope that the dissemination of this knowledge will spread.

The Farmer Treatment Protocol

A typical treatment regimen consists of a variety of worm-killing substances at every meal, plus anthelminthic medication.

Breakfast includes coffee (caffeine and boron) with cloves and cinnamon brewed in. (Cinnamic acid in cinnamon is an analog of phenylalanine, while cinnamaldehyde is also fairly close. They probably penetrate the blood-brain barrier, while curcumin from turmeric does not [Wikipedia].) Hash brown potatoes are a good source of boron (Modern Nutrition in Health and Disease, MNHD), and turmeric, cinnamon, black pepper, garlic, and onions should be added to complement them nicely, and they should be cooked/sauteed in olive oil (anthelminthics). An egg or two for protein can be stirred in just before removing from the heat. A glass of real, unfiltered apple juice would be a good source of additional boron (MNHD). Alternatively, a glass of chocolate protein powder in milk can suffice for a quick breakfast. Chocolate is a source of boron and copper, as well as caffeine (MNHD).

Alternatively, a smoothie with pineapple, apples, oranges, and berries with celery and carrots plus protein powder and perhaps some cinnamon and whatever other additions might be tasty is a good anti-worm concoction. Soy protein powder is better than whey.

For lunch, chicken or tuna salad sandwich can be very healthy: Take one large can of chicken or tuna, add two spoonsful of mayonnaise, two spoonsful of sweet relish, one teaspoon of cinnamon, one teaspoon of turmeric, some raisins for copper and flavor, one finely chopped stalk of celery (anthelminthic), three or so tablespoons of honey (a powerful anti-parasitic, antibacterial substance), a few tablespoons of olive oil, a few cloves of garlic, a sprinkling of black pepper and some salt to taste. Stir. I then eat it spread on a single piece of whole wheat toast. The phytates/phytic acid in whole wheat can bind zinc and copper and prevent them from being absorbed, so don't forget your copper supplements, and vitamins and other minerals! You might also try using some hummus dip, which is increasingly easy to find in grocery stores, these days, as an after lunch, before dinner snack. It has chickpeas (presumably a good source of copper) and garlic (anthelminthic) in it, and it feels very cleansing. You might also want to snack on some celery (anthelminthic) and maybe some raisins (good source of copper). Peanut butter (good source of copper) on crackers is also tasty. A carrot is a good anthelminthic addition.

For dinner, load up some pasta and chicken with as much garlic as you can stand—a few cloves, add olive oil, add on a sprinkling of some turmeric and cinnamon with some black pepper, sprinkle with parmesan cheese. If you like fish, a homemade tartar sauce can be made with a few tablespoons of sweet relish, a similar amount of mayonnaise, and some garlic, turmeric, cinnamon, and black pepper to taste. Approximately an hour or so after dinner, take praziquantel (25mg/kg) dissolved in alcohol such as peach liqueur or whisky or something about an hour or so before bedtime. In the beginning, alcohol-praziquantel induces sleep quite dramatically, and it will make arising in the morning quite difficult, so it is best to take it the night before mornings when one can sleep late. For me (I weigh 100 kg), I take one tablespoon of praziquantel (2.9 g) dissolved in approximately one tablespoon of peach liqueur. Taking alcohol-praziquantel paralyzes the worms when they are full of toxic anti-worm substances such as garlic, turmeric, and cinnamon, etc.—or at least that's my operating assumption, one that has produced very good results so far, curing Parkinson's, Multiple

Sclerosis, ALS, GERD, gluten allergy, Chronic Fatigue Syndrome, osteoporosis, dementia, etc. This protocol provides an invigorating attack on the parasites of the eyes and the rest of the body. Because the parasites are most active at night, taking praziquantel in the morning seems to be less effective than taking it before bedtime. As evidence of this, the parasites often cause rectal itching at night as they crawl out of the rectum for some reason, possibly reproduction. Many people experience this but don't understand the significance—the itching is caused by parasitic worms. (This is also the reason that some dogs will drag their buttocks across your carpeting—it looks cute at first, until you realize that they are trying to relieve the itching that is caused by worms, and they are wiping those strongyloides worms all over your carpeting, and the worms can penetrate the skin.) Additionally, applying diatomaceous earth (DE) to the rectum at night helps to kill the parasites that crawl out of the rectum, apparently by drying them out, and possibly by lacerating them. DE is available from pool supply stores and some health food stores.

In Summation

Blindness is but one of many diseases that plague humanity, these days. Of these many diseases, parasitic worm infection, quite likely with *Strongyloides stercoralis,* easily explains a great many of them. I personally have had more than 40 diseases, some of them quite debilitating, but, mercifully, only a few relatively minor diseases that have affected my eyes, despite the fact that eye diseases and blindness are in my family history. As my health began to decline seriously, some people told me that 1 was just getting old, but that argument seemed specious to me, as I have subsequently proved by dramatically reversing my decline. I think that reasonable investigators, in evaluating the causes of blindness, must apply Occam's Razor, the concept that the simplest solution is the most likely one, to this situation. Clearly, one simple source of causation that explains 40 or more diseases (there are supposed to be about 80 autoimmune diseases, and I now seriously doubt that autoimmune diseases really exist) is much, much more likely than having 40 or more ridiculously complex and convoluted explanations involving mystery genes and phantom viruses with random periods of onset and various other esoteric phenomena. In addition, everyone above a certain age shows signs of copper deficiency that is best explained by parasitic infection.

The question arises: How is it that no one has noticed these parasites before? I don't have a fully satisfactory answer for that. Strongyloides worms were first identified by Dr. Louis Normand, a Frenchman, in 1876, more than 130 years ago *(Wikipedia)!* The worms are tiny, no more than two millimeters long, and they are only rarely found in the stool, apparently, despite the fact that they are known to crawl out of the rectum at night, causing nocturnal itching. It seems that they must be very good at hiding. Perhaps they are just the right size to escape observation—too small to be seen with the naked eye, and too large to be seen easily with a microscope. Perhaps they are best visualized with a special stain that is rarely used. Maybe it's just cultural arrogance, and they are perfectly visible if only some trained person would look for them. In any case, the circumstantial evidence of their presence is overwhelming, including the fact that anthelminthic therapy causes shrinkage of both the eyeballs and also grotesquely-swollen, tumor-like lymph nodes, shrinking them in quantities that can be measured in pounds. Plus, I've seen them in my eyes.

To sum up, I believe that blindness should be relatively easy to cure in many cases, having finally recognized the cause and a suitable treatment. Certainly my protocol for curing blindness is a great deal simpler, easier, and cheaper than many other proposals that are currently being investigated, these days, such as installing robots in the eyes and electrodes in the brain to stimulate the brain directly and some of the other brilliant-but-misguided-and-perhaps-excessively-complicated solutions that are being proposed. Once confirmed and implemented by others, my protocol will cure and prevent the majority of the causes of blindness for many people forever.

It's time that America woke up to the significance of parasitic worm infections. As always, please vote for me for the Greenberg Prize and the Nobel Prize in Physiology or Medicine, if you are an eligible voter. Also, a MacArthur fellowship grant or a Gruber award to sustain me for a while would be nice. Thanks—Robert S Farmer, MD.

Talent hits a target that no one else can hit. Genius hits a target that no one else can see. —Arthur Schopenhauer

SECTION 3

Miscellaneous Ramblings

When I saw that nothing resulted from [doctors'] practice but killing and laming, I determined to abandon such a miserable art and seek truth elsewhere ...

All the universities and all the ancient writers put together have less talent than my arse. —Paracelsus

"Relatively few viruses have any connection with the production of neoplasms," *Rous said. But bulldoggish and unwilling to capitulate, Rous lambasted the idea that cancer could be caused by something inherent to the cells, such as a genetic mutation. "A favorite explanation has been that oncogenes cause alterations in the genes of the cells of the body, somatic mutations, as these are termed; but numerous facts, when taken together, decisively exclude this supposition"* ... *The somatic mutation theory [is] dead. [1966]* — S. Mukherjee, about Nobel laureate Peyton Rous.

Great spirits have always encountered opposition from mediocre minds. — Albert Einstein

Chapter 12

Smoking

All things are poison, while nothing is without poison; the dose makes a thing a poison or not. — Paracelsus

I just want to make a few points here quickly.

The first points are: 1) that smoking has been shown to improve Parkinson's disease 2) virtually all schizophrenics smoke 3) smoking helps many people to deal with anxiety 4) smoking helps to treat emphysema, although some people may not understand that just yet. 5) The officially-recognized oldest-person-who-ever-lived (Jeanne Calment) was a smoker, thus proving that smoking is not necessarily detrimental to health; she smoked approximately two cigarettes daily for 97 years, until her death at age 122½ or so. The dose makes the poison.

All of the above facts are easily explained by parasitic worm infection, and inhibition of it by smoking and the chemicals in tobacco smoke. Killing or inhibiting the worms by smoking reduces symptoms of parasitic disease.

In emphysema/COPD especially, smoking seems like the opposite of what should be done, at least on a superficial level of logic. And yet, people with emphysema seem usually to get worse when they attempt to quit smoking, a finding that is usually explained by saying that the patient is just too addicted to quit, because the patient's body is just too accustomed to smoking. This is a ridiculous, insane, specious rationalization! Emphysema is caused by worms, not by smoking; smoking is the treatment, although there are better treatments available that have fewer side effects and that are not noxious to bystanders. In a study, only about half of patients with COPD were smokers (see my article on the lungs).

I also believe that the body's response to parasitic worms is the main cause of obesity. This is why people often find that smoking helps to keep them thin—the smoking is killing or inhibiting the worms that would otherwise have to be suppressed by fat production. Interestingly, researchers have found that obese people have *increased* metabolisms, when they expected to find *decreased* metabolisms; this strongly suggests parasitic worms, now, doesn't it? Obesity tends to be a strange combination of internal emaciation combined with weight gain due to fat. It's worms!

Here is how worm infections progress:

At first, worms cause emaciation because of the intense competition for nutrients. Emaciation is also caused by the body's exhausting and intensive efforts to fight the parasites. As the parasites fail to be eliminated, the body's supply of copper, especially, and other minerals, such as selenium, become depleted. As the copper and other minerals are depleted, the worms proliferate, causing the body to secrete fat to try to sequester the infectious organisms. These events cause obesity, loose skin, and dark bags under the eyes, as well as various diseases, including heart disease, cancer, and diabetes mellitus. As the person gains weight, he/she teeters on the edge of collapse for years, perhaps while beginning to smoke in an effort to lose weight or to feel better. Signs of aging appear—the exact same signs as copper deficiency: gray hair, wrinkled skin, spinal degeneration from collagen destruction, dementia (forgetfulness), macular degeneration, joint pain, etc. Eventually people die of "old age," which is synonymous with parasite-induced copper deficiency, in many cases.

Parasite-induced copper deficiency can happen at any age. This is why some kids are overweight. However, obesity usually strikes in middle age. As the mineral contents of factory-farm fields continue to be continuously and gradually depleted, more and more people are becoming obese as their copper and boron levels fall as a result of deficiencies in their diets, especially as health-food fanatics harass people into quitting coffee drinking and smoking—habits that fight worm infection, albeit not perfectly.

Many people who smoke have been falsely made to feel guilty for allegedly causing their own health problems and eventual deaths by smoking. It's bad enough that these many people have had their health destroyed by doctors' self-righteous misdiagnoses, but then they are made to feel guilty for using the behaviors that somewhat mitigate their suffering. And the last ideas that these patients are thinking about as they die is that they caused their own deaths—and that is totally false and wrong! This is just falsely adding insult to injury!!! Smoking is the only treatment that these patients have had, because their doctors' treatments are pointed in entirely the wrong directions and are therefore almost totally useless! And often worse than useless—homicidal! And while society pressures people to quit smoking, smoking cessation is causing many more deaths as the worms proliferate, laughing all the way to the worm maternity ward, reproducing like crazy—it's just insane!!

A few years ago, television network news anchor Peter Jennings died of lung cancer, which, he was told, was caused by having smoked 20 years earlier. That is absolutely stupid and insane!!! The reason that he died was because he had worms and he *stopped* smoking! (And a few other things.) There is no way that smoking 20 years earlier would lead to lung cancer. I've read repeatedly that cancer risk for smokers who quit is essentially the same as a non-smoker's after five years. It's just another example of doctors killing their patients due to incompetence.

Also, I keep seeing commercials from [censored], featuring some poor guy who claims that he got lung cancer from second-hand smoke. I will tell you right now that the cancerous effect of second-hand smoke is a lie; the concept is idiotic; and second-hand smoke is not a cause of cancer. (But I don't like it, either.) Did the person who was the (primary) source of the smoke get cancer? Probably not, because no more than 20% of heavy smokers ever get lung cancer, according to researchers. These TV commercials are intensely offensive to me, because these lies are all about gaining money and power, and they are killing people just to increase their funding. Don't fall for it—remember that the oldest person who ever lived was a smoker. Also, I endured years of second-hand smoke—which I hated—from my mother, and my lungs are fine, for my age. Of course, my case doesn't prove this concept conclusively, but I'm sure that the evidence on the other side is even weaker.

So, if you want to quit smoking, kill your worms first. And please do quit smoking (or at least reduce it), because I hate smoking—but at least I understand it, now.

As always, please vote for me for the Nobel Prize in physiology or medicine, if you are an eligible voter.—Robert S. Farmer, MD

Chapter 13

Kidneys, Diabetes, and Urinary Tract Problems

Diabetes resembles fasting [i.e. starvation], especially regarding the responses of the liver, muscle cells, and adipose tissues. — James W. Anderson

The scalpel is the greatest proof of the failure of medicine. — Dr. Juvenal Urbino, a character in *Love in the Time of Cholera*, by Nobel Laureate Gabriel García Márquez

Kidney disease is more common than ever before in the United States. As someone who had multiple sclerosis and diabetes mellitus, I know how easy it is to develop bladder/kidney/urinary problems, because bladder problems are part of the diagnosis of multiple sclerosis. I also technically had diabetes mellitus type 2 for a very brief period, but it seems to have cleared up. The fact that I've had about 50 or more different diseases illustrates a point that should be obvious: many "different" diseases are caused by the same infection, and the distinctions between them are essentially meaningless.

If you ask a modern doctor to define diabetes mellitus, he or she will say that it is an autoimmune attack against the pancreas that causes high blood sugar which damages organs. If you were to check the original description written by the person who is credited with naming diabetes— Aretaeus of Cappadocia, circa A.D. 130—you would see that he described it as "a liquefaction of flesh and bones into urine" (Porter, p. 71). This is a substantially different description! It's a description that certainly sounds exactly like parasitic worm disease. How to explain the discrepancies between this ancient interpretation and the modern one? The answer is that the modern view of diabetes is just a small part of a much larger whole, and doctors refuse to consider this whole, preferring to treat every sign and symptom separately—probably because doctors just don't have the brainpower to consider that what they have been taught is wrong, or at least dramatically incomplete. Or maybe because they make more money by treating every sign and symptom separately. Either way, doctors are very, very narrow-minded.

If we look at the first (the modern) description of disease, we see that there are several problems with it. The first problem is that the concept of autoimmune disease is a fallacy. The second problem is that the attack is not just against the pancreas, it is against the entire body. The third problem is that the attack is infectious, not from or by the body. The fourth problem is that high blood sugar (glucose) has never been shown to damage organs. But do you know what high blood sugar does do? It feeds parasitic worms! And it probably makes them hyperactive. The only useful feature of the common understanding of diabetes is that it allows people to slow down the disease process by starving the worms somewhat using insulin. Perhaps the worms need higher levels of blood sugar to function well, when compared to mammals. Whatever the mechanism, diabetes mellitus is definitely caused by parasitic worms, and it is recognized as being associated with many other diseases that are also caused by worms, although this causation is not yet recognized. It is not genetic, and people who claim that it is genetic are promoting bad science.

I see a lot of commercials on late night TV promoting urinary catheters, and I am just shocked that doctors are so stupid that they let their patients decline to the point where they have to shove a tube up their urethras whenever they need to pee, and that people have to live this way for decades. What more proof do you need about how utterly stupid doctors are? If they would just open a book once in a while, they could avoid this atrocious situation, but doctors are so lazy and passive and lacking even the slightest desire to innovate or experiment—I can't even describe how utterly useless—worse than useless—most modern doctors are today, much less explain how they became this way. Perhaps it's due to the rampant nepotism in medical schools. I had a teacher once who said that he had a good friend who was on the admissions committee of the University of [censored] Medical School. The story that I got was that there were three criteria for admission, in rank order: 1) Are the parents doctors? Because they would be likely to make extra donations. 2) Are the parents at least millionaires? Same reasoning. 3) Is the applicant a good student? Of course, after the slots were filled up using the first two (one) criteria (criterion), there usually wasn't room for good-but-poor students. The person who allegedly said this is reported to have resigned in protest, which undoubtedly accomplished nothing. Now, to return from my digression:

I've had a lot of diseases that generally affect the kidneys and bladder, such as multiple sclerosis, diabetes mellitus, Parkinson's disease, chronic back infection. In chapter 50 alone of the *Merck Manual 17th Edition*, there are approximately 15 diseases that have so much overlap that it is kind of amazing that they are even considered to be separate diseases, but they are clearly all due to parasitic worm infection; there are around 300 chapters in the *Merck Manual*, and many of them are filled with idiopathic (of unknown origin) diseases. Many idiopathic diseases are caused by parasitic worms that doctors refuse to acknowledge. (The *Merck Manual* acknowledges these worms, but mostly in a separate context. Interestingly it also reports that worms are known to cause seizures, yet neurologists ignore this obvious cause of epilepsy. Why? Stupidity is the only reason that I can think of. Case in point: Ben Carson, famous neurosurgeon—and not the sharpest tool in the shed—who would prefer to cut children's brains out, leaving them horribly crippled, rather than treat them with cheap anti-worm drugs!)

For several years, I was having difficulty holding my urine for medium-to-long periods, and when I finally did urinate, it was just a tiny amount, and I never really felt like it all came out. So the bottom line is that I had to urinate too frequently, and the amounts each time were inadequate and unsatisfying. And you knew that this was coming—I cured it by treating for parasitic worms!

The last time that I had a physical and a urine test, my urinary system was in tip-top shape, as good as ever. You can cure your urinary problems, too, but you need to understand that killing your worms is a long-term project, a lifestyle choice. Despite what your quack doctor tells you, your problems are not due to old age, and they *can* be reversed—it just takes time and motivation. Also, my penis works fine in every way—no erectile dysfunction here!

Earlier, in the introduction, I pointed out that doctors' common understanding of kidney disease and hypertension is backwards, and I promised to explain the discrepancy. Most, perhaps all, doctors think that hypertension (high blood pressure) damages the kidney. Maybe it does, but that is not the initial event in kidney disease, and the initiating event is the most important event, so let's discuss that.

You see, doctors use a model of hypertension that has a pump, tubing, and a fluid; the heart is the pump, the arteries and veins are the tubing, and the blood is the fluid. Using this model, doctors decide whether to strengthen or weaken the heart's contractions, to speed them up or slow them down; they decide whether to dilate or constrict the blood vessels; they decide whether the blood is too thick or too thin. And that's it! Did you notice anything missing from this model that all doctors are taught? No? How about a control switch, like a thermostat, except for blood pressure— perhaps a dimmer switch? Well, this model doesn't account for any internal control device—there's no power switch, no dimmer switch, no thermostat. This is odd, because doctors know that it's the kidneys that control blood pressure (mainly), and they use a variety of drugs to control the chemicals that the kidneys secrete (e.g. ACE-inhibitors and angiotensinogen receptor blockers)—so doctors often try to control blood pressure without taking into account the chemicals that the kidney secretes (mainly erythropoietin, angiotensin, and renin), possibly because cardiology and nephrology (the study of the kidney) are separate disciplines, with separate billing practices, and doctors don't really interact with each other, anymore. The kidney increases blood pressure by secreting these chemicals, and it reduces blood pressure by ceasing its production of these chemicals. So if the kidney is the main source of blood-pressure-controlling chemicals, then how does blood pressure get too high? The answer is that blood pressure gets too high—possibly high enough to do damage—because the kidney secretes the chemicals that it needs to raise the pressure to the point where the filter can filter the blood properly. But if there is something interfering with the filtering mechanism, then the kidneys secrete more of the chemicals to increase blood pressure even higher, to the point at which they can again resume making urine—and then you get high blood pressure. So the point here is that blood pressure is low unless there is something that is interfering with blood filtering and urine production; if something interferes with that process, the kidneys raise the blood pressure. So you see, high blood pressure is a response to interference with the filtration system. What can interfere with the filtration system? Well, parasitic worms interfere with kidney function; the right kidney is directly adjacent to the liver, where this infection begins, and the worms can travel in the blood, anyway. Now, what we have is a hypothesis that needs to be tested—that disease leads to hypertension more than the other way around; and, when we test the hypothesis by treating for parasitic worms, we find that hypertension decreases, kidney function improves, bladder function improves, and urination normalizes. Therefore, hypertension is not the primary cause of kidney damage; mostly, kidney damage is the cause of hypertension. Then a positive-feedback loop is set up, and then hypertension does eventually lead to even more kidney damage. The important point here, though, is that the initiating event was a chronic parasitic worm infection that started the chain of kidney damage. I think that the main cause of kidney damage according to the alleged authorities is diabetes mellitus, and because parasitic worms cause "both diseases" (really just one disease), then treating kidney patients will make kidney dialysis and kidney transplants obsolete.

In 1990, Joseph Murray won a Nobel Prize for pioneering kidney transplants in 1954. He had to wait 36 years to get that prize, and I am now declaring that the age of kidney transplantation (and organ transplants in general) is essentially over. Organ transplants are really an admission of the defeat of medicine. If you can eliminate, or just reduce, the disease (i.e. worm infection) in your kidney or heart or liver, then you won't need a transplant. In fact, recognition of the extent of parasitic worm infections in humans will lower the need for many other transplants, too, I'm sure. For the first time in history, humans can really exert effective control over their health, unlike any other previous time that has ever existed.

Chapter 14

Madness and Addictions

I cannot here avoid giving my most decided suffrage in favor of the moral qualities of maniacs. I have no where met ... fonder husbands, more affectionate parents than in the lunatic asylum, during their intervals of calmness and reason. — Philippe Pinel, *Memoir on Madness* (quoted from *Kill or Cure*)

I'm just going to touch lightly on something that should by now be very obvious: brain, mental, and many psychological and neurological problems are caused by parasitic worms; I feel 100% certain that this includes epilepsy, although I have not had the chance to prove this belief yet. Maybe Ben Carson isn't smart enough to figure it out, but it should be obvious to everyone else by now. To me, autism and Alzheimer's disease are essentially the same disease, just at different life stages. The beta-amyloid plaques that everyone obsesses on so much seem to be merely normal scar tissue. The deficiencies of neurotransmitters such as dopamine, norepinephrine, epinephrine, and serotonin are all explained perfectly by parasitic worm infection, copper deficiency, and various associated deficiencies of such things as Vitamin C, tyrosine, tryptophan, and some other miscellaneous-but-essential minerals such as magnesium, etc. The schizophrenic people to whom I've spoken clearly have many health problems that are virtually all explainable by parasitic worms and copper deficiency, etc. It is very simple.

To the extent that mental illnesses are inherited, most of them are inherited due to congenital infection by worms rather than by genes, and of course, some mental illnesses are just defective learned patterns of behavior.

Maybe I should stress this extra hard, just so everyone is absolutely clear: copper is essential to form the above-mentioned neurotransmitters, and without copper, you go crazy and die! Copper is also required to form the myelin sheath on larger nerves, which is essential to normal nerve function. When you see some crazy person on the street, remember that the only difference between him/her and you is a treatable infection that doctors were too stupid to treat and some nutritional deficiencies that doctors were too stupid to notice and/or address properly. That, combined with a job lost due to health problems that were beyond the individual's control, and many of us are only a few days away from being crazy and homeless ourselves. The majority of the crazy and homeless are not really all that different from the rest of us. (As a side note, I am appalled that hospitals still feed their patients things such as white bread and sodas made with high-fructose corn syrup. Doctors have no useful understanding of nutrition, at all.)

Many schizophrenics and homeless people have drug addictions, and that shouldn't be surprising, because heroin, as one example, is a pain-killer; if you have a horrendously painful infection that your doctor refuses to treat (e.g. worms), you have little choice but to try to get pain killers from somewhere else—hence the heroin addiction "crisis," which I blame the medical profession for completely. Many people treat their worms with alcohol, usually after their doctors repeatedly fail to do anything useful, and thus become labeled alcoholics, and then they die—and it could all be prevented by one competent doctor. The heroin crisis is caused directly by the collective stupidity of doctors. Ivermectin paste is cheaper than heroin, and it's probably the safest drug ever made, yet ignorant doctors recoil in horror when I suggest that they prescribe ivermectin without a lab report that says "worms." Why? Stupidity! Ivermectin is apparently the first drug since 1957 to win a Nobel Prize (along with artemisinin, another anti-worm herb/drug [developed by You You Tu]), so one might suspect that it is a very good drug. (Actually, William C. Campbell and Satoshi Omura won the Nobel Prize for discovering/modifying ivermectin.) It is also listed by the World Health Organization as an essential medicine that should be easily and freely available in every country in the world. I predict that treating for parasitic worms will virtually eliminate the heroin crisis.

When I was younger and did a lot of rock climbing, I frequently put myself in situations that were both stressful and extremely dangerous—situations such as, "one toe-or-finger slip and you fall 1,000 feet onto a pile of rocks without even bouncing off the wall." Yes, that kind of dangerous. To be a good rock climber, one has to have confidence—but realistic confidence. Rock climbers who are overconfident don't live very long. Consequently, when I say that I'm confident about being able to cure the heroin crisis, you should know that I have the kind of confidence that I have literally bet my life on many, many times. If you want to help addicts, you're supposed to have all kinds of licenses, permits, releases, certificates, government approved and inspected facilities, etc., etc., etc. But all you really need to make a difference is to start out with some ivermectin paste, and I would bet my life that it will make a difference in almost all cases. In contrast, methadone doesn't treat anything; it's just a profitable (for some people) way to manage addiction more safely—an approach that is not altogether invalid. However, pain-related addictions can be *cured* rather than just managed, but only if you attempt to treat the cause, rather than just using bandage solutions.

Also, the mass murders by people who are clearly schizophrenic (James Holmes, Aurora theater; Adam Lanza, Sandy Hook School; Jared Loughner, the shooter of Congresswoman Gabrielle Giffords in Arizona; and others) obviously have extreme parasitic worm infections at the roots of their causes. James Holmes in particular had been treated for quite a few years for schizophrenia, but of course his doctors were clueless. Adam Lanza was scary-emaciated at the time of his crimes, and he had extreme photosensitivity, as suggested by the way that he blacked out the windows of his basement—a clear case of worms. Psychiatrists aren't expected to know about real medicine, apparently; it seems that they are taught just to give drugs that mask or control signs and symptoms. I think that the psychiatric profession might not even recognize infection as a cause of mental illness. This needs to change. Also, it would be nice if police no longer just automatically shot mentally ill people to death in the streets, etc.—a crazy man with a knife who is 100 feet away from anyone is clearly not a threat who needs to be shot to death immediately. People need to realize that mental illness is just an infection that happens to normal people; once that happens, maybe the public will be less afraid of and more sympathetic toward the mentally ill.

Autism, schizophrenia, bipolar disorder (manic depression), epilepsy, Tourette's syndrome, and many other mental/psychological/psychiatric/

neurological disorders can be treated and cured just by treating for parasitic worm infection. Doctors just need to start pulling their heads out of their asses.

BIBLIOGRAPHY

1   Parker, Steve; Kill or Cure; DK Publishing, London. 235.

Chapter 15

Cancer, Sunburn, and Vampires

[The data] clearly emphasize that conventional chemotherapy of advanced invasive and metastatic disease has failed to effect major reductions in the mortality rates . . . new approaches to the control of cancer are critically needed . . . It is therefore essential that we now reevaluate some of our basic assumptions about the nature of cancer and how we approach the problem of treatment or prevention of this disease. — Michael B. Sporn[1]

I don't have any *certified* personal experience with cancer, but it's obvious to me that a great deal of cancer is caused by parasitic organisms—most-likely worms—especially breast, colorectal, lymphatic, leukemic, gynecologic, lung, and undoubtedly others. Worms are known to travel in the lymphatic system, so this claim really requires no big stretch of the imagination—everyone knows that lymph nodes are one of the primary concerns in breast cancer, as one example. The closest that I came to cancer personally was having grotesquely swollen lymph nodes that were never biopsied. I suspect that they would have been labeled as lymphoma, but that isn't really strong enough proof to claim that I've cured cancer. I didn't have health insurance, so I wasn't going to waste money on a biopsy when my treatments were making progress.

I also had a suspicious growth on my nose that a plastic surgeon told me was probably a basal cell carcinoma, but, again, I never got a formal diagnosis, due to financial reasons. I controlled this growth primarily using selenium supplements—whenever my nose would start to swell on the left side, I would take selenium, and this controlled the problem for years—my nose would shrink back to its beautiful, perfectly formed size. Finally, after years of anti-worm treatment, the growth shrunk completely and it has not returned, so far. I have a feeling that it is still there, though, waiting for my immune system to weaken so it can strike again.

Sunburn and Melanoma

The conventional wisdom says that the number of sunburns that one has in early life is a risk factor for developing melanoma, a deadly form of skin cancer. Being a blue-eyed, blond kid, I used to get sunburns fairly regularly, but I have no fear of developing skin cancer, because the conventional wisdom is clearly wrong.

Conventional thinking says that the damage from sunburn is cumulative, so the more sunburns you have had in the past, the more likely you are to get skin cancer. It's easy to see how a non-scientist would reach this conclusion; it's trickier to explain how doctors can be so ignorant. The fact is that the number of sunburns that you have had in the past is an indication of how well your melanin-producing cells are working. Typically, what happens is that you get sunburn because you are not producing enough melanin to absorb the sunlight that hits your skin. This is in part due to genetics, but it also has to do with whether you have enough copper—and some other things—in your body, because copper deficiency is the first problem that is likely to cause sunburn. As you get older, you may have noticed that a lot of your friends or grandparents at the nursing home develop ghastly white skin that prevents them from going outside due to sensitivity to sunlight— their skin burns more easily as they age, instantly upon exposure to the sun. This is because melanin production requires copper. Copper becomes deficient due to chronic parasitic worm infection because copper is required for immune system function (e.g. copper-zinc superoxide dismutase), as well as for detoxifying various chemicals in the liver and elsewhere (e.g. P-450 liver enzymes). Copper is also required to make/repair collagen, nerves, hormones, and quite a few other things, but it is not that prevalent in the diet or in factory-farmed foods, so people tend to become copper-deficient as they age, and then their condition becomes falsely labeled as "age-related," when in fact it should be considered to be a result of chronic infection and the resultant mineral deficiencies. Researchers have made false attributions and confounded their variables in their studies of these phenomena. I can say this because I have reversed the normal pattern of age-related sun-sensitivity, thereby demonstrating and proving that skin sensitivity to sunlight is not strictly age-related, nor invariably linear in progression.

For about 30 years, I had to wear long sleeve shirts, hats, and sunglasses every day (and I do mean *every* day!) due to sensitivity of my skin and eyes to sunlight. (My eyes were also sensitive to fluorescent lights and bright indoor lights.) This sensitivity was exacerbated by taking doxycycline (a tetracycline antibiotic drug) for several years, which increases sun sensitivity as a side effect, probably by depleting copper from the body. Now, having reduced my worm burden considerably and having taken copper supplements for about 10 years, I haven't had to wear hats, sunglasses, or long-sleeve shirts for several years now, although I still wear hats on really intensely bright days. This summer, I will be working a little bit on my tan again, which has never really been very dark. For me, having been a creature of darkness by necessity for so many years, my ability to tolerate sunlight is still quite astonishing, and I now delight in going shirtless at midday.

It's important to have exposure to sunlight in order to make Vitamin D, so living in the dark and/or slathering on sunscreen are bad ideas. Also, sunscreen is said to be a leading factor in the destruction of the coral reefs in oceans worldwide, so please stop using sunscreen! Put on a shirt or hat, instead. Use gloves. Take your copper supplements. Copper deficiency (and possibly worms) is probably a factor in several sun-sensitivity diseases, especially porphyrias, and also epidermolysis bullosa, leukoderma, phenylketonuria, and possibly even xeroderma pigmentosum and some forms of albinism, among others. Regardless of your official diagnosis, if you have sun-sensitive skin, it is most likely due to copper deficiency. Never take Vitamin C without copper, otherwise copper may be flushed out of the body. Also, it has been reported that high fructose corn syrup binds copper and removes it from the body, so if you've been waiting for just one more reason to start avoiding high fructose corn syrup, there it is. Tyrosine (an amino acid) is also required for melanin production; deficiencies of tyrosine are likely to be the second-or-third most important deficiency leading to sunburn. It is found in many proteins, such as soy protein, or you can buy supplements separately.

Vampires

As I alluded to above, I used to be a vampire. Well, maybe not a vampire in the strictest sense, but I certainly understand how the idea of vampires must have come into existence. I know because I too used to be a creature of darkness.

As I have said over and over again, everybody has worms, and everyone experiences increasing copper deficiency as they age, thus they have what appears to be "age-related" sun sensitivity. I've experienced this much more than most people, and as a result, I often tried to avoid going outside before sunset or thereabouts. This was quite inconvenient for someone whose hobbies are rock climbing and white water kayaking, so I would often plan my activities so as to maximize my time in the shade, and to finish up around sunset.

As anyone who spends much time outside around sunset probably knows, bats like to come out from their hiding places about an hour before sunset. I used to spend hours sitting on the porch of my cabin in the Caribbean when I was a medical student down there, watching my cat get up onto the porch railing, and, standing up straight on her legs, reaching up into the encroaching darkness to try to catch the bats who seemed to delight in tormenting her, until she lost her balance, fell halfway off the railing, and had to catch herself and then claw her way back up to do it all again, until the bats became bored and wandered away. The bats were always too smart to get too close, staying just out of reach of her claws.

With this in mind, the myth of vampires seems fairly obvious. Some of the characteristics of vampires include fear of sunlight; pale, white skin; a desire to drink blood; and the ability to change between human and bat form. Vampires often were believed to live in castles, and wealth is a factor that is correlated fairly well with long life; chronic disease gets worse the longer one has it, so old people are more likely to have chronic diseases than are young people. So then, the myth of vampires undoubtedly went something like this: A rich old man who lives in a castle has a chronic disease (worms) that leads to copper deficiency and thus pale skin and sun sensitivity, so he doesn't come outside until sunset, about the same time when bats come out to feed. The villagers would notice that they didn't see the old guy until the bats came out, and they made a fallacious assumption that the man could change into a bat and vice versa. It's reasonable to assume that somewhere along the line, at least one of those old guys experimented and found that drinking blood made him feel better, because the blood of a young animal or person would contain relatively high levels of minerals such as copper and iron; the man wouldn't know why drinking blood made him feel better, but at some point he got caught drinking blood, and the villagers had another reason to associate him with vampire bats, which drink blood. Then, the man and the bats would both go back to their lairs around the same time in the morning. Some embellishments were added, such as a religion-based fear of the silver cross and needing to be killed with a wooden stake through the heart, and there you have it—the vampire myth started with a chronic parasitic worm infection and the copper deficiency and skin photosensitivity that it caused.

While I'm talking about vampires, I'll use this opportunity to go off on a tangent and say that parasitic worms also almost-certainly had a hand in the fall of the House of Romanov—the czars of Russia—and the rise of communism. The young prince Alexei, son of Nicholas, had hemophilia, a bleeding disorder (said to be an X-linked genetic disease) associated with problems of Factor VIII; Factor VIII requires copper to function. Therefore, it is likely that Alexei's bad health and hemophilia—which were factors in the rise of the Russian Revolution, because he was seen as a weak and sickly heir to the throne—were possibly caused or exacerbated by parasitic worms and copper deficiency. Parasitic worms have undoubtedly altered the course of human history many, many times throughout the ages.

When pondering the cause of melanoma, it's important to note that, while melanomas are regarded as being caused by the sun, they often appear in places that get little or no sun exposure, such as the perineum (between the legs), on the buttocks, and on the bottoms of the feet. This pretty much proves that the sun is not really the cause of melanoma, and I've reversed so-called age spots or liver spots by treating for worms and taking mineral supplements including moderately high levels of copper and other minerals, including selenium, along with vitamins. I'm not sure of the exact mechanism, but it certainly seems that melanoma is certainly related to copper deficiency in some way (seeing as how copper is required for normal melanocytes, or melanin-producing cells). I'm certain that copper deficiency is related to other cancers, too, because most people with cancer tend to be either overweight or underweight and have loose, pale, lumpy skin—signs of worms and copper deficiency.

The best example of copper deficiency in cancer would be acute lymphocytic (lymphoblastic) leukemia (ALL), which has three main characteristics: 1) pale skin 2) extreme bleeding 3) extremely high levels of defective white blood cells. This disease is usually defined almost exclusively by the high white cell count, but white cells have nothing to do with bleeding, which is the cause of death, so focusing on white cells mainly is pointless. I've already explained pale skin; the extreme bleeding can be caused by inhibition of Factors V (five) and VIII due to copper deficiency; and the high white cell count is due to an infection that may or may not be recognized; the cells undoubtedly fail to mature properly due to the copper deficiency, thus feeding into a positive feedback loop that creates huge numbers of white blood cells that are trapped at an immature level of differentiation due to copper deficiency, thus never triggering the stop mechanism. In acute promyelocytic leukemia, defective, cancerous white cells mature into normal macrophages when treated with Vitamin D, a hormone which almost certainly requires copper in order to be formed in the skin when it is exposed to sunlight. I've studied the history of cancer, and I never saw that anyone pursued an infectious approach to cancer, other than William B. Coley, who was pushed aside for reasons of money, politics, and power when radiation "therapy" was invented around 1900 or so. Dr. Coley found that an infection called erysipelas sometimes cured cancers, suggesting that cancer was due to some type of infectious organism, which doctors of the time just rejected for no good reason and because of their obsession with those shiny, newfangled x-ray machines. It is just one more example of how irrational doctors have been throughout the ages.

We live in an age of science, but science has not eliminated fantasies about health; the stigmas of sickness, the moral meanings of medicine continue. Previous centuries wove stories around leprosy, plague, tuberculosis, and so on, thereby creating terror, guilt, and stigma. But the modern age created similar taboos about cancer ('the big C') as untreatable, fatal, and psychogenic, the product of the so-called cancer 'personality', the self that eats itself away through frustration and repressed anger. Therapy was hindered and suffering multiplied. . . . In important respects, science itself has been the vehicle for the proliferation of health fantasies. — Roy Porter

In 1927, Johannes Fibiger was retroactively awarded a Nobel Prize for the year 1926, for showing that a type of parasitic worm could cause cancer in mice. His work was criticized after his death, when he could no longer defend himself, when it was alleged that the cancer was actually caused by Vitamin A deficiency and irritation, either separately or together. Later, his work was included as evidence for chronic irritation as a cause of cancer, but not for the parasitic worms. It seems to me though, that the people who criticized him for not proving that parasitic worms are the single and true, perfect, unitary, definitive, ultimate and only cause of cancer are criticizing him because he didn't find something that doesn't exist—and that is not fair. Cancer is a multifactorial disease—there will never be only one ultimate cause of cancer; rather, there are several ways to induce cancer, and because irritation has been shown to cause cancer, and because parasitic worms cause irritation, it is only logical that parasitic worms can cause cancer, despite the fact that they are not the only thing that does so. (Similarly, Hulda Regehr Clark has claimed that all cancers are caused by an infection of flukes, a type of microorganism; unfortunately, her argument is weakened by the fact that some of her supporting claims are clearly wrong—most notably her claim that copper [an essential nutrient] needs to be completely eliminated from the body—but we apparently both agree that cancer is an infection.) The irritation leads to chronic deficiencies of vitamins, amino acids, proteins, and minerals, including Vitamin C and copper, and undoubtedly other things, too. I don't know whether Dr. Fibiger claimed to have found the only cause of cancer, or just one cause of cancer. It seems to me that he probably would not have made a claim that he had found the only cause of cancer, because that doesn't seem very scientific, but I shouldn't speculate on that. Nevertheless, it seems to me that Dr. Fibiger's critics really missed the point—that chronic inflammation leads to cancer, and parasitic worms are one cause of chronic inflammation; therefore, parasitic worms are one—perhaps even the main—cause of cancer, and might be a cause of Vitamin A deficiency, as well. But physicians have never been the sharpest tools in the shed, have they? If they had just listened to Dr. Fibiger, or Dr. Coley, perhaps cancer would have been cured 100 years ago.

REFERENCE

1   Sporn, M. *Chemoprevention of Cancer.* In: Shils M, Shike M, Ross AC, et al, eds. *Modern Nutrition in Health and Disease, 10th edition.* Lippincott Williams & Wilkins, Philadelphia, 2005. 1280.

Chapter 16

Some Random Famous Celebrities with Worms

Jaguars break into the temple and drink to the dregs what is in the sacrificial pitchers; this is repeated over and over again; finally, it can be calculated in advance, and it becomes part of the ceremony. — Franz Kafka

I'm using some celebrities to illustrate signs and symptoms of chronic parasitic worm infection. This is not meant to embarrass or blame anyone, because we all have this same infection to varying degrees. In the case of people who are dead, I hope that we can learn from their experiences; in the cases of people who are still alive, in addition to being able to learn from them, my message is: I can help you. So there it is. Remember: You can lead a horse to water, but you can't make her drink.

Michael Jackson (musician) died from parasitic worms and the drug overdose that he received from his doctor as a result of his extreme worm-caused insomnia. He had insomnia, vitiligo (patches of white skin), emaciation, light sensitivity of the eyes and skin. A classic case of parasitic worms and copper deficiency with a drug overdose finale. Also, the reason that his nose contracted after plastic surgery was undoubtedly due to parasitic worms in the sinus region. His doctors were terrible—thoroughly incompetent.

Phillip Seymour Hoffman (actor) died as a heroin addict because his doctors were too stupid and incompetent to diagnose his worms. He was obese, with pale, lumpy, skin, and he was in constant pain that his incompetent doctors couldn't or wouldn't treat properly.

William Nealy (kayaker, artist, activist, and author) blew his brains out with a shotgun because his doctors wouldn't or couldn't treat his chronic pain.

Elizabeth Taylor (actress) wasted away slowly as a result of parasitic worms, finally being confined to a wheelchair before she died slowly, obese, crippled, and white as a ghost.

Marilyn Monroe (actress) had the classic, chunky "love handles" on her somewhat large hips/buttocks, and large breasts that are the most typical presentation of parasitic worms in people who produce fat in response to worms. Her face was a bit chubby, but still pretty. She was alleged to have a prescription drug problem, possibly with pain killers. Of course! Her doctors were clueless. She died in her early 30s.

Mickey Rourke's (actor) face has become somewhat misshapen and swollen as a result of parasitic worms migrating into those areas that have been repeatedly injured as a result of being punched in the face innumerable times.

Bill Murray (actor) also has these deforming swellings in his face (see above). Worms love injured tissues, and the face and cheeks are filled with lymphatic ducts.

Robert DeNiro (actor) is just looking ill and what most people would consider to be old, but I could help him reverse his decline.

Kirsty Alley (actress) has had weight problems and large lipomas (fatty tumors) that are typical of parasitic worms.

Carrie Fisher's (actress) numerous problems with drugs, weight gain, etc., and her early death were clearly due to parasitic worms.

Elton John (musician) is clearly loaded with parasitic worms—pale, lumpy, obese skin.

George Michael (singer, age 53) had numerous problems with drugs and medical problems that are easily explainable using parasitic worms.

Bono's (singer in U2) health problems are clearly due to parasitic worms—glaucoma, osteoporosis, visual light sensitivity, etc. I could help you, Bono!

Keith Richards (musician, Rolling Stones) is probably still alive because he seems to smoke nonstop; he looks thoroughly emaciated due to parasitic worms, but he just keeps on going and going. All of those drugs have probably stoned his worms senseless. But they will get him, eventually.

Peter Jennings (network news anchor, 50s) was told by his incompetent doctors that his fatal lung cancer was caused by having smoked 20 years earlier. Nonsense! Smoking is a treatment of last resort for parasitic worms, after doctors throw up their hands in puzzlement and defeat, because they are absolutely befuddled by a disease that literally everyone has! (Gee! I wonder what it could be? Could it be the same thing that literally everyone else has, you morons?!) If he hadn't quit smoking, he might still be alive. Tobacco has been used medicinally for thousands of years, although, of course, it is not a perfect drug, and cigarettes are filled with non-tobacco impurities and contaminants.

Stephen Colbert (comedian, The Late Show) has often commented that he is unsatisfied with his aging appearance and horribly, horribly sagging face. He is younger than I am, but I look younger than he does. I could help him.

Nicole Kidman (actress) is looking scary-white and emaciated, these days, when she's not wearing lots of makeup. It's worms.

Malcom Gladwell (author) has the appearance of emaciation due to parasitic worms.

Paulina Porizkova (retired model) and her husband Rick Ocasek (musician) have the emaciated appearance of a shared parasitic worm infection.

Dr. Oz (TV doctor) is younger than I am but looks much older. He has so far [censored].

Karl Malden (actor, e.g. *The Streets of San Francisco*) had a grotesquely swollen nose that is typically and falsely associated with alcoholism. Actually, part of his problem was that he didn't drink alcohol—doing so would have helped to control the worms that were taking over his nose (as they tried to do in my nose, too). Worms are the leading cause of sinusitis, but doctors are clueless, inappropriately prescribing immunosuppressive drugs that eventually kill their patients, instead.

W.C. Fields (author and humorist) — same as Karl Malden, above, except that W.C. liked alcohol and cigars, I seem to recall, which help to control worm infection.

Tiger Woods (golfer) definitely needs my help: four back surgeries?! Come on, now! His clueless doctors are milking this cow dry. Plus, his sexual addiction — I could definitely help him.

Sarah Silverman (comedienne) had a serious throat abscess that was probably parasitic worms. She also looks a bit emaciated, and she has ganglion cysts on her wrists, a sure sign of worms.

Heather Locklear's (actress) health has taken a steep downward turn, obviously due to worms. Considering that she was the epitome of sexy health, remember that "if it could happen to her, it could happen to anyone."

Roseanne Barr (Arnold?) (comedienne/actress) has always had weight problems. I imagine that she must have a lot of pain, too. I could help her. She totally has worms.

Jennifer Lawrence (actress) famously discussed her gastrointestinal problems on David Letterman's show. Her doctors are useless; she obviously has an intestinal worm problem. I can help you, Jen!

Speaking of which, David Letterman is so old that he could really use some anti-worm treatments. All old people are dangerously full of worms.

Oprah (interviewer, actress, media mogul) has had constant weight problems. It's worms.

George Clooney (actor) is apparently the same age that I am, but looks much older. He could use my help, too.

Jon Stewart (comedian) recently wondered publicly on *The Late Show* why he looks old. It's worms!

Ashley Graham (plus-size model) has an enormous bottom due to chronic infection with worms. Obesity is not just an alternative lifestyle — it's an infection!

Kathy Griffin (comedienne) has deep lines on her face. People think that these are age lines, but they are due to infection.

Anderson Cooper (news reporter) is ghostly white most-likely due to a copper deficiency, most-likely due to chronic worm infection.

Michael J. Fox (actor). Parkinson's disease is caused by parasitic worms.

Richard Simmons (exercise coach/celebrity) disappeared from public view in the fairly recent past, at the usual age when parasites tend to take over the human body. Unfortunately, he has most-likely discovered that exercise is not enough to maintain health.

Sarah Hyland (actress, *Modern Family*) has recently been diagnosed with Lupus or some similar, ridiculous, bogus disease. The steroids that she is taking are allowing her worms to proliferate, causing her cheeks to swell alarmingly. Her doctors are killing her; they have no idea what they are doing.

Jack Nicholson (actor) has fairly recently left the public sphere as his abilities deteriorate and his worm infection approaches its predictable climax.

And last but not least, Chuck Lorre, (TV producer) has used his one second of air time at the end of his shows to talk about how little fun it is to grow old and sick. I could definitely reverse his problems to a very great extent. The problems that he describes are textbook-perfect descriptions of parasitic worm infection.

This is just a somewhat random list of famous people who have, or have had, clear signs and symptoms of parasitic worm disease, and is not meant to embarrass or harm anyone. Every person who dies of "old age" has actually died of parasitic worms. People shrink in height; they develop curvatures of the spine, such as hunchback; they get osteoporosis and bed sores and emaciation and obesity that are reversible if doctors would ever make the correct diagnosis and initiate the correct treatment. It turns out that we are not at the top of the food chain, after all.

Chapter 17

Treatment Notes, The Simple Version

The brutal truth is that the human body is a wonderfully evolved machine, and medicine rarely does as well as nature. — William Bynum

Fixing your health is conceptually simple, but requires high levels of motivation and tireless commitment to healthier living. It is not just a medical treatment but a lifestyle choice that you will need to adhere to for the rest of your life. It's not hard; you can eat whatever you like—but you'll have to eat a lot less of some things and more of some other things that perhaps you are not naturally inclined to like. I enjoy my diet, so it's not hard for me, but if you're accustomed to a diet of fried cheese sticks, potato chips, and beer, then change might be hard for you.

Before we get started, I need to educate you about the Jarisch-Herxheimer reaction (JHR). As the worms die, they release toxins. Some of the toxins are chemicals, but bacteria (apparently gram negative) are also released, which may require antibacterial therapy, such as with a second-generation cephalosporin antibiotic (I recommend cefuroxime). There will be sleepiness, strong nerve pain, gastrointestinal problems such as heartburn and cramps, blurry vision, fatigue, malaise, morning paralysis, etc. These symptoms will last for about 8-12 hours, and then you'll feel a little bit better than you did before. Depending on how committed you are, you will definitely see measurable results within 3 months, and you will almost-certainly feel better much sooner. You will need to experiment to see how much difficulty you have getting up the next day, in order to determine how often you can do the treatments. Remember that this is a lifestyle, not just a temporary treatment regimen. Also remember that in order to feel better, first you have to feel worse. In fact, I believe that the JHR is the main reason for the common hangover—these two are almost (but not totally) synonymous. People experience the pain of the hangover and think it is completely bad, without understanding that the reason that they keep seeking the experience over and over again is because there is a hidden benefit—a dying-off of parasites. See if you can get some pain killers, and use aspirin liberally. Aspirin is an irreversible inhibitor of cyclooxygenase, unlike any other NSAID—all of the others are reversible, which, generally speaking, is not as good. You want a pain killer that can't be inactivated by the pain enzymes—so aspirin is better than ibuprofen or naproxen, etc. If you have stomach problems with aspirin, they should improve as you kill your worms and your health improves.

An Overview

Step One: Kill your worms. This almost certainly will require the use of drugs, in addition to alcohol, herbs, and a few other things.

Step Two: Strengthen your body and immune system. Chronic infection with parasitic worms lowers immune resistance, making you susceptible to a variety of opportunistic infections. Use vitamins and minerals. Minerals are often neglected. Take a multi-mineral supplement. Take extra copper (5-15 mg/day) and selenium (200-600 mcg/day). Magnesium, potassium, calcium, sodium, zinc, iodine (from kelp), are also very important.

Step Three: The use of herbs can help both to kill worms and strengthen the body. Wormwood and Black Walnut is a common combination that is a good start. Milk thistle is also popular. Some herbal cleanses are sold for cleansing the body and/or colon and almost certainly affect worms. You could also consult an herbalist. Herbs may give you diarrhea, but fear of diarrhea is greatly exaggerated—mostly, it just means that you're losing toxins in your stool. Unless it's so severe that you feel it making you weak, don't worry about it—just replace the fluids and minerals in your diet, and you'll be fine.

Step Four: Plant foods contain natural pesticides (i.e. phytochemicals); meat does not have natural pesticides, because animals use antibodies that die after the animal dies. Worms love meat, just like you—that's why they infect you, to eat your muscles, just as you eat the muscles of cattle, birds, etc. So eat plenty of vegetables, fish, beans, nuts, and whole grains. Keep your meat intake low. Fruits, because of the sugars, are not as good as vegetables. Worms often love fruits, but everything in moderation. Corn does not seem to have anti-worm properties. Pineapple is probably the best fruit, because it is loaded with enzymes.

More details:

Step One: Two drugs that I have used successfully are ivermectin and praziquantel. I buy ivermectin paste from horse-related websites (such as horse.com and statelinetack.com). I like it because it is easily available, it has a tolerable taste, it's cheap (about $2 per tube), it's easy to use, and it's effective. Any generic brand should be fine. It will be labeled "not for use in humans" or something like that. It comes in a syringe that you squirt in the mouth; the plunger is marked in pounds or sometimes kilograms, and it has an adjustable locking collar. I have taken almost 1400 doses of ivermectin for horses with no problems. This is almost 300 tubes, or allegedly enough to treat (but not very thoroughly) a herd of almost 300 horses. As an experiment, I've taken up to five times the recommended dose (a full tube) at one time (I don't recommend that, though). Ivermectin may be the safest drug ever made. I often wash it down with alcohol, to try to improve its penetration into the central nervous system. Its inventors/discoverers won a Nobel Prize in 2015; ivermectin is apparently the first drug since 1957 to win a Nobel Prize.

Praziquantel is more powerful than ivermectin, but the side effects are more powerful, too. I dissolve it in an alcoholic beverage, preferably 80 proof (40 percent alcohol [ethanol]) or more. It smells and tastes horrible. I buy it (I use Aqua-Prazi™) from a fish aquarium supply place, and I buy it in the 100 gram size, which costs around $120. For detailed instructions, see the other chapter on Treatment Notes.

Either one of these drugs are available in forms for people, but the price is outrageous, around 100 times more, and you'll need a prescription that your doctor almost certainly won't give you, because doctors behave as automatons that don't function without input from a lab printout that gives a definite numerical value of something, which in this case can't be measured. The human drugs are probably still available only from the Centers for Disease Control, even though the World Health Organization has called them both "essential medicines" that should be freely, cheaply, and easily available in every country in the world. The difficulty in getting them is probably due to the drug industry suppressing cheap and affordable treatments by lobbying the government and the medical profession to limit competition.

Because these drugs cause a Jarisch-Herxheimer reaction (JHR) that includes sleep induction, I usually take them one hour before bed, or just after dinner. (JHR is a release of toxins as infectious organisms die. It is not a drug side effect. It gets much easier to tolerate over time.) Also, the worms are nocturnal, so it's best to hit them when they are active, at night. Your sleep will improve dramatically, but perhaps not right away. I wash the drugs down with food. Other side effects include restless sleep (at first), itching (and itching and restless sleep!), rash, hot flashes, diarrhea, edema, fever, blurred vision, malaise, nausea, and vomiting. The side effects from these drugs are inconvenient, not life-threatening, but I urge you to look them up yourself on the internet or at a library if you tend to react badly to any drugs. There may be other drugs that you could use, but for now these are the best, the cheapest, the safest, and the easiest to get.

Alcoholic Beverages

Other ways to kill your worms include the use of alcoholic beverages. I consider the use of alcohol to be essential. If you're an alcoholic, it's because subconsciously your body recognizes that alcohol kills or inhibits your worms; the problem is that many people don't know when to quit or slow down, even when the alcohol stops providing relief. Also, it's good to rotate different beverages to maximize effectiveness. In the past (and still now, for virtually everyone, because of stupid doctors), there has been no other treatment for worms, except perhaps smoking, so people just drank alcohol until they died, because <u>they couldn't get either adequate treatment or even adequate pain relief from their doctors</u> or anybody else, despite the fact that ivermectin has been on the market since the 1970s or maybe 1980s.

The use of alcohol conforms to the concept of the Therapeutic Window: too little is useless, too much is harmful, but somewhere in between is an amount that is just right (helpful). You can also think of it as the Goldilocks principle. More technically, this is also illustrated by the inverted U-graph, which Malcolm Gladwell discussed in his book, *David and Goliath*. On the left side of the graph, health improves as alcohol intake increases up to a point; next, further increases in alcohol consumption provide no additional benefit up to some other poorly-defined point, then; finally, additional alcohol causes a decrease in health. Beverages such as <u>red wine, brandy, gin, absinthe, and non-corn-based whiskies</u> are excellent for controlling esophagitis (inflammation of the esophagus—when swallowed very slowly), and are very useful for local treatment of asthma, as well as being good for general health, in moderation. (They are definitely better than corticosteroid inhalers—which should never be used—and better over the long-term than beta-2 agonists [for asthma], as well. However, some alcohols such as beer or vodka may provide little or no benefit.) On average, I keep my alcohol intake down to around 2-8 drinks per day (this varies), but this should be based on a dosage-per-weight scale (I weigh about 250 lbs.); this works out to a maximum of 1.6 alcoholic beverages per 50 pounds of body weight. (A drink is considered to be 5 ounces of wine or one ounce [or shot glass] of distilled spirits.) Not all alcohol is created equal: red wine is much more tolerable than whisky or absinthe, for example. I mostly drink my alcohol in the evenings and about one hour before bed, with perhaps a swig or two of whisky or gin after lying down in bed; in order to maximize the benefits, I try to absorb as much as I can through my gums and cheeks by holding the alcohol in my mouth for a few minutes—absorbing it this way delays metabolic breakdown by the liver. This provides more benefits than gulping the liquor down quickly, which causes the booze to be metabolized more rapidly by the liver. It also helps to treat gum disease (which is caused by worms), although it can result in inflammation of the gums temporarily. I consider the withholding of wine from children to be child abuse, because it is so important to health and has such a long record of safety. (Laws against the consumption of alcohol by minors lead to more health problems and more binge drinking when they become 21. Teaching responsibility would probably be a better way to handle the approach to alcohol.) When infants and children suffer from gastrointestinal cramps (e.g. colic), you can bet that it is due to worms. Doctors don't have a clue. Also, I've found that brandy is excellent for brushing my dog's teeth. Seeing as how it is so helpful for my gums and teeth, I know that it is also great for her teeth, too.

It's important to note that the use of alcohol can lower levels of vitamins in the body, especially B vitamins, most famously thiamine (B-1); so take extra vitamins and minerals when using alcohol as therapy. B vitamins are extremely safe, and the recommended amounts tend to be far too low, so don't be afraid to take five times the recommended amounts or so. Alfalfa tablets are a good source of nutrition.

Also, as a side note, I believe that Fetal Alcohol Syndrome is actually due to parasitic worms, and probably not alcohol. If it were due to alcohol, children would get better after birth, when they are no longer exposed to alcohol—but they don't; they continue to do badly, even long after birth, despite the fact that alcohol can be metabolized in about 8 hours. Therefore, the problem is still inside their bodies, suggesting a chronic infection. Alcohol is a confounding variable that distracts from the true cause of their problems—parasitic worms. Worm infection is the reason that the mothers felt compelled to use alcohol in the first place, despite the severe social stigma and legal consequences of doing so while pregnant. Whether alcohol consumption actually causes additional harm is an open question, for me; I doubt whether anyone has asked that question before, let alone investigated it objectively.

The reason that people become addicted to substances is usually due to parasitic worms. For instance, many people are addicted to coffee, tea, tobacco, marijuana, and heroin. These (except heroin) are loaded with natural pesticides such as caffeine and nicotine. I encourage people to try other kinds of tea that may have more anti-worm activity, such as comfrey, sage, foxglove, and a variety of others that are sometimes available from your local health-food store, or which you might be able to grow in your garden. Use these with caution, as they can be very powerful.

<u>I'm sure that many, perhaps all, drug addicts are addicted to pain killers because of the pain of parasitic worms. If you want to cure heroin addiction, treat parasitic worms first.</u> I haven't had the opportunity to prove this yet, but it is obvious.

Beans are much healthier than meat, because they have phytochemicals that undoubtedly act as natural pesticides. <u>Garlic and onions should be considered to be essential eating.</u> Carrots, celery, and tomatoes (tomato sauce is extremely healthy) are also very good. Cabbage is also very powerful, but coleslaw made with corn syrup should be avoided. (Coleslaw made with just a little sugar/sucrose is fine.) Any kind of concentrated fructose syrup is probably unhealthy (possibly excepting honey). (Entry of fructose, unlike glucose, into the cell is not regulated, so the cells can absorb too much fructose, which feeds worms.)

The essential starter kit of herbs/spices that are great for the gastrointestinal tract as well as for the body generally include:

Turmeric – ½ teaspoon
Black Pepper – ½ teaspoon
Cinnamon – ½ teaspoon
Granulated Garlic – ½ teaspoon—or fresh raw garlic
Olive oil

I pour them into my mouth and wash them down with red wine, tea, or olive oil, or perhaps coffee or vegetable juice or something. Vegetable juice is very helpful, in general, too.

Pets

I still use the aforementioned spices every so often, and I have given a little bit of these to my dog virtually every day for about 8 years, and her health is very much better than average. (I sprinkle a reasonable and proportional amount into her food bowl.) I've reduced her intake of olive oil lately, because she doesn't seem to like it anymore, although I thought that she did, in the beginning. I try to cook up a large pot of spicy beans and vegetables every so often, and she loves this stew mixed in with her fish-based dry food, instead of the canned food that I usually mix into her fish-based dry food. She loves spaghetti sauce, and this is very good for dogs, as it is for people.

There's a badly-done study somewhere that says that garlic is bad for dogs, but this is ridiculous. Too much of anything is harmful for any organism; I've given my dog the first four spices virtually every day for eight years, and she is the healthiest 9-year-old dog whom you could ever see. She hardly ever walks—she mostly just sprints from scent to scent.

My cats have an automatic feeder that dispenses dry food, and before inserting the food, I put it in a large canister with the above-mentioned spices, with the addition of one small scoop of soy protein powder, and I shake it up to distribute the spices and soy powder onto the nuggets. At first, my kitties really hated it, giving me dirty looks and going on a hunger strike and meowing indignantly and pitifully. Eventually, though, they got used to it. I give them canned food every few days, too. The spice regimen has definitely helped them to lose fat and to be healthier, and now they seem to prefer it, maybe.

I use small, appropriate amounts of ivermectin on my dog and my cats, and it is very good for them. If you were to go to a veterinarian, he/she would probably be willing to sell you Iverheart™ at a considerable mark-up over generic ivermectin—about $40 for six pills, plus the cost of an office visit. The last time when I talked to a vet about getting ivermectin for my cats, he wanted to sell me some other, less-safe drugs that, when combined with the office visit that he would require just to get the drugs, would end up costing about $150 per cat, versus about 10 cents for a cat-sized dose of ivermectin paste for horses. Sure, just let me run out to my Rolls-Royce limousine to see if my chauffeur brought my checkbook...

Also, as mentioned earlier, brushing my dog's teeth using approximately one-third of a shot of brandy to dip the toothbrush in seems to benefit her teeth as it does for me, too.

Diets

There are a lot of diet books on the market, but they are all (but one) pretty much the same—eat vegetables, grains, beans, fruit in moderation, fish and poultry, meat and dairy in moderation, alcohol in moderation, use healthy fats in moderation; avoid excessive amounts of fried food. There's no way that I'm going to count calories in any but the most approximate sense—it's way too much trouble for me. When you're healthy, you don't feel compelled to overeat. Still, you should exert some self-discipline, especially as your caloric needs decrease as you kill your worms.

I used to drink a lot of milk, up to a gallon on some days, and probably close to one-half gallon per day on average. This was too much. I believe that I did it because I craved the galactose (milk sugar) and phosphate for energy. Unfortunately, milk is very alkaline, and worms love alkaline environments. Your goal should be to try to stay slightly acidic—worms don't like that. Now I limit myself to one or two large glasses of milk per day, and I don't drink it every day, anymore, either. Use diluted apple cider vinegar, in a glass of juice or something before a meal to increase acidity. Rinse the teeth immediately.

Speaking of which: There is a common misperception that the reason that people drink sodas and become addicted to them is because of the sugar in them. However, this does not explain the popularity of sugar-free sodas, so this belief is clearly wrong or incomplete. The real reason that people drink sodas is because they are the highest source of phosphate in the diet, and phosphate is pure energy. (Biochemistry reference: Think: adenosine triphosphate, ATP.) Phosphate is added to sodas to make the carbonation. (To demonstrate, take a spoonful of Rumford™ baking powder [not soda] and add it to some fruit juice—instant soda!) Milk is also high in phosphate, which is undoubtedly part of the reason that I drank so much of it—I was starving due to parasitic worm infection, even while getting fat!

Another note about dieting: When you kill off your worms, you will lose fat, but not necessarily weight. The reason is because dieting normally results in the loss of fat, muscle, and bone; however, when you starve yourself, the worms shift more to eating your muscles and bones because you have reduced the supply of food. Therefore, after killing your worms, you will lose less muscle and bone while dieting—the proportion of fat loss will increase. Worms are the major cause of muscle loss (doctors measure muscle breakdown as creatinine in the blood, but they don't know what causes it—worms!), so when you kill worms, you will maintain more muscle even while dieting. My appetite is now about 25% of what it used to be, and I eat about 75% less than I used to eat. This means that 75% of the food that I used to eat was going just to support my worms! Now, I maintain strength much, much better between workouts than I used to, and, of course, I feel much better, too.

Hot peppers and hot spices in general have anti-worm (anthelminthic) properties. So use more spices, herbs, peppers, etc. Eat whole foods such as whole (not white) wheat and brown (not white) rice. Wheat is a lot more nutritious than rice, and alfalfa tablets are an excellent supplement in addition to soy protein powder. Whey protein powder probably feeds worms, whereas soy has phytochemicals that seem to inhibit them, so I recommend soy, not whey.

Apple cider vinegar is good for the health, but you absolutely have to dilute it in some juice or water! I discovered the hard way that pouring it into your mouth undiluted can result in a potentially-fatal asthmatic reaction, even if you've never had asthma before. It was very frightening! Dilute!

Also, organic apple cider vinegar seems to be much better than the regular ultra-cheap brands that you are undoubtedly attracted to on the basis of price; organic apple cider vinegar is much more expensive, but it is probably worth the extra money. Bragg's is a common brand, and they seem to have built up a cult-like following due to the numerous benefits that they claim for their vinegar. Remember that the extra cost is just a drop in the bucket when you figure up how much your illness has cost you over the years. I mix 1-3 tablespoons of vinegar in V-8™ juice with a little orange juice and soy protein powder in the mornings. You can also add cinnamon and nutmeg, etc. Bragg's has a website: www.Bragg.com.

Also, wheat germ oil seems to be a very good addition to the diet. It is somewhat expensive at health food stores, but it is available for horses in a wheat germ oil/soybean oil mixture that is much more affordable, although then you have to buy it in a gallon jug that needs to be refrigerated, which is kind of a pain.

Many people find that they get indigestion from spicy foods, or wine, or some idiosyncratic food, ("I can't eat that!") and the reason is generally that your worms are reacting to it, probably always because they don't like it. As you get healthier, you'll find that you can eat any spicy food—even wasabi—and it's no big deal, well, except maybe for wasabi, but you'll be able to eat wasabi without melting into a sniveling puddle of tears, at least. You should force yourself to eat those foods that give you indigestion—using good judgment, of course—because those are the foods that the worms hate the most. After a while, spicy food will become a non-issue for you, and you can sneer derisively with your smug sense of superiority when you hear some unfortunate person say, "I can't eat that!"

Emergency note:

I just saw on TV that some wrinkled, gray-haired, half-senile old fool of a doctor has decided that pain killers cause or contribute to heart disease. This confused, addle-brained quack has seriously misplaced his attribution—he is completely unable to distinguish between cause and effect. Pain killers are not the *cause* of heart disease, they are the *response to* heart disease, which is probably almost-always caused by parasitic worms—which tend to be very painful—that doctors are too stupid to recognize. According to this so-called doctor, the people with the worst heart disease use the most pain killers, so therefore pain killers must be the cause of the heart disease. There is an expression here that applies: *Post hoc, ergo propter hoc*, which is a definitive fallacy well-known to people who study logic. It means, "after this, therefore on account of it," and it is a definitively false conclusion. He is saying that their heart disease gets worse after taking pain killers. This is like saying: "I see that marathon runners sweat and breathe heavily, and I notice that they drink a lot of water during the race; therefore drinking water causes sweating and heavy breathing!" It is a ridiculous conclusion, reached by a twit.

A lot of people say that they have an allergy to aspirin, so they take ibuprofen, naproxen, or some other things to avoid stomach pain. I think a little lesson in pain killers is appropriate here. Aspirin was invented and brought to market in 1899 (or so) by the Bayer Corporation. It is often abbreviated ASA, which stands for acetyl salicylic acid; it is a synthetic version of salicin and salicylic acid, which are natural chemicals that were originally found in the bark of the white willow tree (Salix alba—probably no relation to Jessica Alba) and/or meadowsweet, although it is also found in several other trees and herbs. So these plants didn't make this chemical to make humans happy; they developed these chemicals for their own needs—salicylic acid is a natural pesticide, and aspirin is a modified version of this natural pesticide. When people experience stomach pains apparently because of aspirin, it is most-likely because of the Jarisch-Herxheimer reaction (JHR). Reducing your worm infection will make your stomach almost impervious to any kind of irritation. Aspirin is my favorite non-prescription pain-killer. Unlike any other NSAIDs, the effect of aspirin on the pain enzyme cyclooxygenase is irreversible; in contrast, ibuprofen, as one example, wears off in about 8 hours, so not only has the body made more pain enzyme in those 8 hours to replace the inactivated enzyme, but now you have to deal again with the same pain enzymes that you just inactivated 8 hours ago! So now you have twice as many pain enzymes! It's like that scene in Men in Black, when the alien gets his head blown off and then regrows it in about ten seconds, complaining about how inconvenient that was. If using pain killers is analogous to shooting the head off an alien, then with ibuprofen, the head grows back in about 8 hours, but with aspirin, the alien stays dead.

Doctors seem to be under a lot of government pressure these days to avoid prescribing opiate pain killers, because of the foolish/insane belief that opiates have some kind of magically addictive power. (Addicts are in pain, dumbass! [I would say to critics.] What do you think causes pain, hmmm?) However, until the government stops harassing doctors and drug users, there's a prescription drug named Tramadol that is not on the opiates list and provides good pain relief without gastrointestinal problems. However, the best long-term treatment is to kill your worms; then you won't need pain killers, anymore. I used to be horribly aversive to relatively mild kinds of hot peppers, as one example, and now I can chomp them right down without a thought. It's a dramatic change that has been achieved by killing my worms. I used to drink a tremendous amount of milk to ease my stomach pain, as another example, having been diagnosed with a stomach ulcer starting at age 7 or so, yet now I have the proverbial cast iron stomach.

I used to live in southeastern Kentucky—which is regarded as the opiate addiction capital of the U.S.—as part of my medical training, and I can tell you that their health problems are largely due to parasitic worms, which doctors have not recognized. I've observed when people are told for the first time that they have diabetes, or when they are falsely told that they will die of emphysema or heart disease if they don't quit smoking immediately. I particularly remember one old lady who told us, "Well, if I have to choose between living longer and continuing to smoke, then I'll just keep smoking." At the time, it seemed to be an astonishing thing to say, but, of course, neither I nor the doctor whom I was studying under had ever smoked, so we were both ignorant of what smokers go through. Now, having smoked a bit for experimental purposes, it is obvious to me that smoking inhibits worms pretty strongly, but unfortunately not for very long. It really can produce a remarkable feeling. As Paracelsus said, "Everything is poison. Only the dose makes a thing not a poison." Smoking is kind of a disgusting habit, but it has been falsely vilified by ignorant extremists as the cause of diseases that it most certainly is not the ultimate cause of. Tobacco is one of the best examples and definitions of a confounding variable—in other words, it confuses medical inquiry—it is a false attribution for disease; it is blamed for things that are not its fault. Tobacco has been used as medicine in the Americas for thousands of years, and it was quite popular as medicine after being brought to Europe around 1500, which was a time when few effective treatments for diseases were available. Later, around 1525 or so, Paracelsus introduced laudanum into Europe, which was a revolutionary opiate liquid, and later, in 1632, quinine, in the form of Peruvian bark, was introduced for the treatment of malaria, in particular, and more generally for other things.

If there were a lot of other very effective treatments for diseases, I haven't come across them in my readings, although there were many treatments of indeterminate utility. Today, the war against tobacco is really more of a religious obsession, like the Spanish Inquisition. Facts don't seem to change the minds of tobacco's enemies, although to be fair, neither side has really understood what was really going on.

Emergency Note 2

I recently saw that some insane quacks have announced that alcohol—even one glass per day—causes breast cancer, or increases the risk, based on some associative study that they did. These morons are unable to distinguish between cause and effect. Alcohol is a partially-effective *response* to cancer— it most certainly is *not the cause* or even *a* cause. Now, beer *might* help cancer to grow, but that is pure speculation. However, condemning all alcohol as bad is simply irrational, illogical, delusional, insane, and stupid. People who drink alcohol moderately live longer than non-drinkers, and that's a fact!

Also, worms seem to tend to gravitate downward in the body: they end up in the wrists and hands, the lower back and pelvis, the scrotum and vaginal labia, and the legs, feet, and ankles. It is important to keep the genital/scrotal area clean, because the worms seem to like to emerge from the body there (and around the rectum, causing rectal itching) to reproduce. Dusting with diatomaceous earth, especially at bedtime, is a good idea. These worms are the reason that people use talcum powder—to reduce the genital itching caused by parasitic worms. The recent lawsuits against the manufacturers of talcum powder for allegedly causing ovarian cancer are almost-certainly a confusion of cause-and-effect. I doubt very much that talcum powder causes cancer. Worms cause cancer, and they also cause moisture of the groin that makes people want to use talcum powder down there.

Emergency Note 3

I recently came across a book called *The Vitamin D Revolution* by Soram Khalsa, M.D., that makes a compelling argument that everyone should get at least 4,000 I.U. (International Units) of Vitamin D every day. I was already very conscious of Vitamin D, but apparently even I was not conscious enough. After taking 50,000 I.U. one time to start, I felt better the next day, with less of the chronic shoulder pain that I haven't been able to get rid of completely, yet. For those who get their levels tested, Vitamin D-3 should be more than 50 ng/ml.

As I have stated elsewhere, it seems virtually certain to me that the formation of Vitamin D (actually a hormone, not a vitamin) in the body requires copper, so it stands to reason that people who are low in Vitamin D have serious parasitic worm infections and the copper deficiencies that those cause. Any vitamin D supplement that you use should be the oil-filled capsules, because, as a fat-soluble vitamin/hormone, it cannot be absorbed in the dry form.

In any event, all of this should give you a good start on your recovery. When you feel like advancing to the next level, the following chapter is a bit more difficult to plow through, but it does have some very good information in it.

Chapter 18

Appendix — Treatment Notes Two, Deeper Thoughts

The following section goes into quite a bit of detail, and is reproduced from an earlier version, with minimal editing. There is some repetition, which I apologize for. There are references to my earlier experiences in anti-worm treatment that I hope will be useful.

Updated: 9/26/12

Before jumping into my treatment notes, I just want to say that I now believe that everyone is born with a worm infection, and that dietary and other habits determine how rapidly or to what extent the worms cause disability. I have also come to the conclusion that worm infection is most-likely the primary determinant of human lifespan, unless a person dies of something else first.

In order to help people achieve a cure from worm infection, I'm going to review my treatment regimen here. I used to not recommend that people undertake to treat themselves without the guidance and supervision of an intelligent doctor, but, as it turns out, the vast majority of doctors are at this point in history far too stupid to be useful, so do what you have to do. First, let me say that taking copper or multi-mineral supplements alone will not be sufficient to effect a cure. Remember that even as you strengthen yourself with the mineral supplements, you may also be strengthening the worms unless you take anti-worm drugs. Remember that strongyloides (as the presumed infection) infections are lifelong unless treated, and the longer that you have been infected, the longer the treatment period will be. There is some repetition in the following notes, due to combining several articles—I apologize for that.

Parkinson's Disease/MS/Anthelminthic Therapy

To effect recovery, it will be necessary to take copper supplements. Copper is essential for the function of neutrophils and other immune system cells, and I suspect that it is essential for the function of antibodies (immunoglobulins), although that has not been proved, that I know of. I also suspect that copper is involved in the hydroxylation reactions that convert cholecalciferol to 1,25 dihydroxycholecalciferol (a.k.a. vitamin D), which is important or even essential in fighting cancer. Several hydroxylation and/or oxidation reactions are also essential to convert cholesterol to hormones such as aldosterone, cortisol, and the sex hormones (estrogens and testosterone); these probably require copper and vitamin C, too.

When replenishing the body's supply of copper with supplements, a theory of mine is that copper, like iron, is absorbed better with vitamin C. (They both have the same atomic charge, sometimes.) I don't know if it's because of the acidity of vitamin C (a.k.a. ascorbic acid), or whether there is something special about vitamin C and absorption. Copper often works with vitamin C throughout the body (lysyl oxidase to cross-link collagen is but one example), so it makes sense that they should be taken together. Taking vitamin C without copper tends to flush copper out of the body, and I'm sure that you know someone who takes vitamin C but has never even heard of taking copper, but it is probably not a good idea to take vitamin C without copper (about 1000 mg vitamin C for every 5-10 mg of copper, I estimate) in order to avoid copper deficiencies.

Copper and zinc often work together, in opposition throughout the body. I have read that the copper: zinc ratio should be about 1:10. I recommend taking more copper, about 10 mg of copper with 50 mg of zinc (or maybe even more copper), because the main problem here is copper deficiency, and apparently not so much zinc deficiency. I am not sure that zinc deficiency is really a problem, and it does compete with copper for absorption (allegedly), so I often don't take it at all, other than combined with multi-vitamins, etc. Take copper with food, or you will throw up! Because some people say that copper and zinc compete for absorption, antagonizing each other, they should be taken at separate times. I do try to take them at different meals, when I take zinc, because I think I feel a difference, but I'm not really sure.

I have recently decided that taking too much copper might help the worms to resist the antibiotics. I noticed that my lymph nodes stopped shrinking and actually grew a little after taking copper supplements. This would make sense if, as I suspect, the worms oxidize the medicine to inactivate it, copper being a major oxidant (although, strangely, it's also an anti-oxidant—you know how you're always taking anti-oxidants?). For people who have been severely debilitated for a long period with Parkinson's/MS/ALS etc., you still need to take copper, but as you get better, you will want to try to strike a balance between filling your own needs for copper while simultaneously starving the worms of the copper that they need or want. Perhaps taking a little copper around noon would be best, because the worms are more active at night. Obviously, this is a pretty tricky balancing act (and it may not be important), but when you finally get to the point where you are functioning normally, you may have enough copper and don't need to take any more. Until you reach the point of normalcy, however, an inevitable amount of experimentation will undoubtedly be required. Hey, this is cutting-edge medicine, here, and working out the kinks takes time! Note: I now think that this is probably not a problem with my latest regimen.

I recently read about some research which determined that <u>fructose binds with copper</u>, flushing it out of the body. OMG!!! This is an unmitigated <u>public health disaster</u>, because high-fructose corn syrup is hidden in almost every food, these days. It's apparently the cheapest and most-popular sweetener in the U.S. by far. If you have Parkinson's/MS/ALS, etc., you definitely need to <u>stay away from concentrated fructose</u> in any form (except maybe fresh fruits, because they are healthy). Considering that <u>copper is essential for insulin function and thyroid function</u>, are you really surprised that the rates of Diabetes Mellitus and obesity are rising out of control? Fructose binding copper makes perfect sense when you evaluate the current state of deteriorating American health—Parkinson's, MS, diabetes mellitus, allergies, are all associated with low copper levels—and worms, I believe. *S. stercoralis* plus high-fructose corn syrup will be proved to be a very deadly combination, no doubt!

Additionally, good nutrition is important. Remember that you've been feeding a bunch of tiny little freeloaders for a very long time, and they have stolen a lot of nutrition from you. Take high doses of multivitamins, take a tyrosine supplement, and take a protein supplement such as whey. With the whey protein powder that I take (I was severely allergic to it for about two years or more, when I had to stop taking it, using soy instead), there are about 700 mg of tyrosine in one small scoop and about 750 mg of phenylalanine (which can be converted to tyrosine, using copper), according to the label. But seeing as how you've been deficient for so long, a little more tyrosine won't hurt you, so throw in that tyrosine supplement, too, (unless you're on a super-tight budget, in which case you can probably get by with taking only protein powder for the tyrosine)!

Here's a quote from *Wikipedia* listing some good food sources of copper:

Foods contribute virtually all of the copper consumed by humans. The best dietary sources include seafood (especially shellfish), organ meats (e.g., liver), whole grains, legumes (e.g., beans and lentils) and chocolate. Nuts, including peanuts and pecans, are especially rich in copper, as are grains such as wheat and rye, and several fruits including lemons and raisins. Other food sources that contain copper include cereals, potatoes, peas, red meat, mushrooms, some dark green leafy vegetables (such as kale), and fruits (coconuts, papaya and apples). Tea, rice and chicken are relatively low in copper, but can provide a reasonable amount of copper when they are consumed in significant amounts.

If you have worms, then you are deficient in all minerals. I am sure that worm-induced mineral deficiency will be shown to be a major cause of what is generally and probably inaccurately called osteoporosis (osteomalacia is more likely). How can I make this claim? Well, my leg bones (tibias) used to feel so weak and painful that I literally worried about breaking my legs by stepping off of curbs. I literally broke my leg once by jumping six inches to the ground while twisting my ankle; I've suffered four broken legs on three occasions as a result of rather minor impacts, despite the popular claim from medical authorities that being athletic, as I was/am, is good protection. Also,

I believe that my mother died of worms, and she broke her thigh (femur) by tripping over a plastic bag full of clothes in her bedroom, which is just freaky...and wrong. If you have "osteoporosis," then taking only vitamin D and calcium, even with estrogen, "just ain't a-gonna help nothin.'" (I doubt that taking estrogen is really all that useful—it probably is either converted to vitamin D or is a poor imitation of vitamin D; in any event, vitamin D can possibly be converted to estrogen, so just take vitamin D, in an oil base. Both vitamin D and estrogen are made from cholesterol.) There are many minerals involved in bone formation, and worms apparently love to eat them, but I am digressing again, and this is a discussion for another time. However, until then, kill your worms, and take multi-mineral supplements, with extra copper, even with iron, even if you are a man. And stay away from those anti-osteoporosis drugs—they destroy the normal architecture of the bones; although they increase bone density, they don't improve bone strength, and they are not the best treatment because they do not address the underlying cause of weak bones—worms!!!

More about Copper

Looking more closely at copper, the U.S. Recommended Daily Allowance is only 0.9 mg/day, while the Estimated Average Requirement is 0.7 mg/day. While authorities used to believe that people consumed abundant levels of copper, more recently experts have decided that few diets have even 2 mg of copper. While the anti-vitamin supplement crowd likes to complain that nutritional supplements are unnecessary and only give Americans the most expensive urine in the world, it is important to remember that the RDA is for healthy people who do not engage in strenuous physical activity. However, many people are clearly not healthy, and they know it, while other people may think that they are healthy, but actually are not.

Therefore, the nutritional needs of many people are increased well above the RDA. There are few if any supplements that are toxic at a few times above the RDA, so it seems reasonable to take up to 5 or maybe 10 times the RDA of most supplements, with the possible exceptions of vitamin A and iron, and maybe a few other things, including zinc. Copper is easily eliminated in the bowel, so taking a dose that is moderately increased above the minimum shouldn't be a problem, except possibly for people with severe liver disease, and they need more copper, anyway. Copper does pass into the kidney and is apparently mostly reabsorbed rather than excreted, and it is possible that high doses might be a problem for kidney dialysis patients, but I really don't know anything about kidney dialysis. I will say, however, that copper is <u>essential</u> for healing and maintaining integrity of tissues and organs, so I think it is implausible that kidney dialysis would be a contraindication to moderate copper therapy. Here's a note about copper toxicity from the USDA website that indicates that copper supplements are really pretty safe at sane levels: (note missing)

When taking copper, it seems likely that it is best to take it with vitamin C; experiments have shown that iron is absorbed better with vitamin C, and because copper often works with vitamin C throughout the body, it makes sense that vitamin C should help with the absorption of copper. It is best to take copper with food, because it tends to cause immediate vomiting when taken on an empty stomach. When taking 5-10 mg of copper, approximately 1000 mg of vitamin C should be about right, I'm estimating.

While many people routinely take vitamin C especially in the winter to ward off colds, it is worth mentioning that taking vitamin C without copper lowers copper levels in the body, effectively flushing copper out of the body. In my opinion, vitamin C should never be taken without copper, for this reason. Other things that tend to inhibit copper absorption in the gastrointestinal tract are zinc and antacids, including anti-heartburn drugs such as proton pump inhibitors and histamine blockers. Likewise, taking more than 3,000 mg of vitamin C per day is probably a bad idea. I've tried megadoses of vitamin C, and I found the results of this experiment to be unsatisfactory, Linus Pauling notwithstanding.

Zinc

Zinc should probably be taken at separate meals from copper, if zinc is needed. Copper and zinc often work together, but they compete for absorption. Milk is very low in copper and tends to lower the body's level of copper, but I love milk, so I keep taking my supplements. Also, I recently read that milk is an excellent source of phosphorus/phosphate, which I have been deficient in, too, apparently, so I have unconsciously been treating my phosphorus deficiency for 35+ years by drinking lots of milk!

Boron

Boron is an essential mineral, classified as a non-metallic element. Boron/borates have been used as medicines and food preservatives for more than 4000 years, and they have been shown to have antibiotic properties, being a part of several antibiotics that are made naturally by myxobacteria (*Modern Nutrition in Health and Disease*). Between 1850 and 1900, boron-containing substances were used to treat epilepsy, which tells me that it must kill or inhibit worms, because I believe that worms are the most-likely cause of epilepsy. From the 1870s until 1904, boron was used to preserve fish, meat, cream, and butter, but it was banned when someone observed that it reduced appetite and caused indigestion (*Modern Nutrition in Health and Disease*). There is quite a bit of evidence that boron helps strengthen bones and increase production of hormones such as testosterone and estrogen. It is considered to be completely safe and effective as an eye wash. I am currently testing boric acid to see how it works for treating my infection, and initial results are positive.

Boron is an essential part of plants, although plants require much more boron as a percentage of their total minerals than people do, for forming their skeletons. In the U.S., the highest concentrations of boron are found in Death Valley, California, which has no outlet to the sea, because it is below sea level. Not too far to the north, there are some ancient volcanoes, the watersheds of which drain into Death Valley. This helps to explain why Death Valley has such a high concentration of boron—because boron is usually brought to the earth's surface by volcanoes. Volcanoes also explain why the highest-ranking source of boron in the average American diet is coffee. Coffee is almost-always grown on the sides of volcanoes, and at least 17 of the 20 countries with the highest coffee exports are countries with volcanoes (see my article on Coffee). Other relatively high sources of boron are apples, milk, beans, and potatoes. Washington State, which has several volcanoes, is, of course, a state with a large apple industry (58% of U.S. apple production—from http://urbanext.illinois.edu/apples/facts.cfm), and the downwind state of Idaho is famous for growing potatoes. Other foods that are high in boron are avocados, dates, prunes, nuts, honey, and wine (*Modern Nutrition in Health and Disease*); the USDA website adds peanuts and peanut butter, grape juice, chocolate powder, pecans, and granola raisin and raisin bran cereals (http://www.nal.usda.gov/fnic/DRI//DRI_Vitamin_A/502-553_150.pdf). It's worth noting that cacao beans, the main ingredient in chocolate, are often grown with or near coffee beans, thus they would presumably share a mineral profile, as seems to be the case, suggesting that coffee may also be a good source of copper.

In a study of boron intake among residents of Germany, Kenya, Mexico, and the U.S., Americans had the lowest boron content, while Mexico had the highest (*Modern Nutrition in Health and Disease*). This should be no big surprise, as Kenya has volcanoes, and Mexico City is completely surrounded by volcanoes. It appears that most Americans are deficient in boron.

You can go to your local vitamin store to buy boron supplements, but the amounts available (2-3 mg) will make supplementation expensive. Boric acid is much cheaper.

In using boric acid as a nutritional supplement, use only medical grade boric acid, which is available at your local drug store for about $5. It is labeled "not for internal use," probably for reasons that have more to do with legal protection and profits of other drugs than actual safety, because my boric acid bottle says that it is 100% boric acid; however, you do want to remember that only very small amounts of boric acid are necessary. One-quarter teaspoon of boric acid is about 175 mg of boric acid, which provides about 31 mg of boron. Although this is much more than the average daily intake and more than what many sources recommend (2-3 mg/day), it is a lot less than what people used to take in the late 1800s. Epilepsy used to be treated with more than 500 mg of "borates", apparently, which equals about 87 mg of boron/day, if boric acid was used; it would be less boron if borax was used; 500 mg of borates is apparently the level at which toxic effects began to appear (*Modern Nutrition in Health and Disease*), but some other toxic effects may occur at lower levels, such as inhibition of sperm in men—however, that could be a good thing, if you know what I mean. It's unclear how much boron is optimal, but I'm starting to think that the RDA is too low, as it often has been shown to be with other essential nutrients.

I have personally taken ½ teaspoon of medical-quality boric acid (labeled "not for internal use") in my mouth for six days, at which point I decided to stop due to headaches and general malaise. This was a very high dose, equivalent to about 3.5 mg/kg of body weight per day, but I was experimenting. The effects that I have experienced by taking boric acid are similar to the effects that I experienced by taking praziquantel, leading me to believe that boron supplements will probably be essential for eradicating worm infection, as it seems to help the function of other minerals in the body, according to various sources. Taking ¼ teaspoon for a few days as an initial dose and then once or twice a week indefinitely is probably safe for most adults. Because boric acid is an acid, but probably not as acidic as Vitamin C (ascorbic acid), you do not want to leave it on your teeth—rinse it off immediately!

Another way you can take boric acid is by heating or boiling one cup of water; turn off the heat and add one teaspoon of boric acid, then stir it in until it dissolves (the melting point of boric acid is 175 degrees). Wait for it to cool. Using a regular eyedropper, each full squirt of the eyedropper equals about 3 mg of boric acid, or about 0.5 mg of boron. In this manner, by squirting one eyedropper of boric acid solution into my cat's mouth, I cured his what-appeared-to-be asthma. However, he was still having prolonged sneezing attacks, so I gave him another dose, and he has not had another major allergic sneezing attack since (I give him some about once a week). I also squirt some onto my dog's food about once a week; it is almost tasteless. I've also noticed that since taking boric acid, I no longer have any tremors when I get very hungry, which I've previously attributed to phosphorus deficiency due to refeeding syndrome. In any event, boron is an essential nutrient, and the evidence that Americans are deficient is pretty overwhelming, and boric acid is perfectly safe in very small quantities; use your head and stop if you experience problems, although you should expect to experience some inflammation as your immune system wakes up after its long sleep to attack your worms and other parasites.

You can look up some very long discussions of minerals including boron and toxicity on the USDA website.

Pharmacological Issues

In looking for medicines to try to test my theories, I discovered that getting anti-worm medicine is fairly difficult through the usual channels. The usual drugs recommended for strongyloidiasis are ivermectin, praziquantel, albendazole, and thiabendazole. Thiabendazole is said to be an older, more toxic drug that is available only by prescription, so it seemed unavailable. Albendazole is a newer version, and I was able to buy it from India over the internet at a reasonable cost. However, it did not seem to help much, even when I took twice the recommended maximum dose, which caused more than half of my hair to fall out and turn gray. (Hair loss isn't that likely at the recommended dose; and my hair grew back in its original color.) I was also considering ivermectin and praziquantel, but these drugs have to be special-ordered, and they would cost about $1,000 for about a week's worth of medicine—prohibitively expensive, about $50,000/year. (Supposedly only one dose is needed, and it's supposed to last for six months, but that claim is just absurd!!!) Then my second cousin, who owns horses, mentioned that she could buy enough ivermectin to treat a horse for about $10. I was skeptical. Surely there must be some horrible, horrible reason why people shouldn't take horse medicine, right? After all, a horse-sized dose is five or six human-sized doses, so how can one account for the enormous discrepancy in price between human medicine and horse medicine?

Currently, the only reason that I can see for the price difference is greed on the part of pharmaceutical companies. Ivermectin is given away for free in Africa. I have now taken enough ivermectin and praziquantel to treat a herd of more than 400 horses [in 2012] (that's about 1300+ human-sized doses), and I have had zero serious side effects over more than two years, approximately. I like Equimax™, because 1) it tastes less bad than the other medicines, and 2) it has a mixture of ivermectin and praziquantel, although the level of praziquantel is too low to be therapeutic. (I should note here that Equimax comes in a syringe, but you do not inject it, you squirt it in your mouth, I mean your horse's mouth, in proportion to your, I mean your horse's, body weight.) In order to achieve therapeutic levels of praziquantel, I buy praziquantel for the treatment of pond fish (e.g. goldfish and koi—for my inner-city koi pond) over the internet from fish-supply companies. (Look for Aqua-Prazi from Aqua Meds or Microbe-Lift from Ecological Laboratories. Webb's Water Garden has been a reliable source. If you buy 100 grams, it costs around $120. You might want to check out their instructions, too, while you're on their websites.) The label says that it is pure praziquantel, but it also says that it is not to be used in fish that will be used for human consumption. I have found praziquantel to be the most useful medicine for treating my condition, which I evaluate in part by examining the size of my grotesquely swollen lymph nodes (a very objective evaluation, but they are sometimes impossible to palpate precisely when they are extremely large), although it may work best in combination with ivermectin, in some cases. I have now taken about two pounds (454 g per lb x 2 lbs=908g) of praziquantel, about 2.9 grams (one tablespoonful) at a time, dissolved in drinking alcohol (I prefer peach liqueur or whiskey), which helps it to penetrate the central nervous system. (The recommended dose is about 25 mg/kg of body weight, three times per day, maximum [2.5g for a 100-kg person, 3 times per day—about 75mg/kg/day]. Because of the intense Jarisch-Herxheimer reactions that make me sleep and then make me feel somewhat paralyzed and exhausted the next morning, I could tolerate only one dose every few days or weeks at first, although the negative side effects diminished as my health improved. I weigh about 100 kg or 220 pounds.) My improvement has been dramatic, losing more than 10 pounds of swollen lymph nodes, and a lot of body fat, and a tremendous amount of pain, while regaining my memory and my mind, along with many other improvements, including decreased tremors and increased exercise capacity. The praziquantel seems to be more effective at night, when I'm sleeping, which is when the worms are said to be most active, strangely enough (explaining my occasional insomnia).

I prefer to treat any infection with at least two or three drugs, so it's a good idea to combine praziquantel with ivermectin, which I have bought for as little as $1.99/horse-sized syringe (for generic ivermectin alone), plus shipping costs (from horse.com). My Parkinson's/MS/ALS has improved despite never having taken any anti-Parkinson's Disease or anti-MS drug, ever. Those drugs mask the symptoms, they don't treat the cause.

More about Praziquantel

Praziquantel is a drug that the World Health Organization lists as an Essential Medicine, meaning that it should be easily and cheaply available in every country in the world. Praziquantel is the drug that has really brought me back from the dead. Fortunately, it is available without a prescription, although this version that I have been using is not labeled for human consumption. Unfortunately, I am not wealthy enough to be exploited by the big pharmaceutical companies, so I feel very fortunate that I have been able to take advantage of this back door route to regaining my health. One has to wonder why human drugs cost about 100 times more than the same drug that is labeled for horses or fish. Personally, I think that the people who run the pharmaceutical companies don't care how many people die, as long as they make money so they can live in mansions and drive expensive cars. I really think that the government has really abdicated its responsibility to protect the American people from greedy exploiters, when it comes to something as fundamental as health care. The praziquantel that I use is labeled for the treatment of pond fish, and is labeled "pure praziquantel," and I have had no problems with it.

In any event, the maximum recommended dose of praziquantel is 25 mg/kg of body weight, up to three times per day, for a total of 75 mg/kg/day. It comes in the form of a powder that needs to be dissolved in drinking alcohol, apparently, or at least doing so helps it to penetrate the central nervous system and probably various other tissues, as well. The highest dose that I've been able to tolerate is up to perhaps around 30 mg/kg/3 times per day for 2 weeks, however, I needed months of smaller doses to work up to that, because of the severe Jarisch-Herxheimer reactions caused by the toxins released when the parasites die. One tablespoon equals 2.9 grams, and there are three teaspoons in one tablespoon. Also, there are 2.2 pounds in one kilogram. Because I weigh about 100 kg (220 lbs), I take about one tablespoon of praziquantel dissolved in about one tablespoon of peach liqueur, which hides the taste more than Triple Sec.

The usual side effects are nausea, headache, vomiting, and intestinal cramping—all fairly mild. Also, it tastes terrible, but it is not the worst drug that I have ever tasted, although it is very close. I may have vomited one time due to high levels of praziquantel, about 12 hours later, but I can't be certain of the actual cause. This is despite having taken 600+ doses, so it is really quite a mild drug, although I do have a hard time moving in the mornings due to a kind of mild paralysis due to the Jarisch-Herxheimer reaction presumably due to the anticholinergic release of toxins as the parasites die, or perhaps due to anticholinergic effects of the drug itself. In general, praziquantel appears to be safer than aspirin, according to my drug reference books. Especially in the beginning of my treatment, I felt compelled to sleep within an hour of taking praziquantel, although the extreme sleepiness diminished somewhat as I cleared more of the infection, and now I can sometimes stay awake after taking a dose, but I can not do so consistently. It is metabolized by liver enzymes CYP 3A4, so it interacts with metronidazole to some extent (see below). As always, proceed with caution.

Ivermectin

Ivermectin works by blocking glutamate-gated chloride channels in the worms, causing paralysis. There's that chloride again, perhaps explaining in part why I have had an allergy to salt (sodium chloride), and why my allergy to salt diminishes as I kill the worms . . .

The ivermectin that I use is marketed for horses. Generic ivermectin can be bought for as little as $1.99 to treat a 1,350 pound horse, from horse.com or statelinetack.com. The human version of the drug costs about 100 times more. The horse ivermectin comes in a syringe that is marked in 100 kg or so increments, and you squirt it into your mouth. There are no real calculations involved, other than possibly converting pounds to kilograms (2.2 lb/kg). If you're crazy rich or have great health insurance, the actual recommended dose is 200 mg/kg of body weight, if your doctor will prescribe the human version of the drug. Supposedly, only one dose is required every six months, but this is clearly dramatically inadequate for the infection that I have. I'm not sure how often it can be taken, but I've taken double doses daily for a week or two, for certain, and it has caused me no problems. As always proceed cautiously. For a detailed history of ivermectin, find/go to Satoshi Ōmura's website.

Albendazole

I bought some albendazole from India over the internet, but it didn't seem to help me a lot. The maximum dose is 400 mg, two times per day. I took twice that amount for a while, and it made my hair turn gray and fall out, but it grew back in its original color after stopping the drug. You should probably stick to the recommended dose.

Etc.

I had several years earlier experienced some relief with mebendazole, but at $35 per pill, I found the price to be prohibitively expensive. My friend Dr. Jerroll Dolphin tells me that it costs about ten cents per pill in Africa. I also experienced some relief with metronidazole, which is not an anti-worm drug, but is more of a general antibiotic that works well in the intestines; it is used to kill amoebas (amoebae) and protozoans (protozoa), among other things. The dose ranges from 250 mg to 750 mg three times per day for up to two weeks, and you should avoid alcohol when taking it. (I seem to recall taking 500 mg/3 times per day.) It also interacts with a variety of other drugs and perhaps a few foods, so it is not nearly as foolproof (i.e. safe) as praziquantel and ivermectin are. Metronidazole affects liver enzymes CYP 2C9, CYP 3A3/4, and CYP 3A5-7, so it does interact with praziquantel (CYP 3A4) a bit, although I have never personally experienced any problems with these interactions, and sometimes I have actually used them with other drugs to try to boost levels of drugs in my system, but as someone who just doesn't understand or use recreational alcohol or drugs, I'm a freak of nature, so I can't say that you and I should use the same set of rules. Also, I know how to boost my liver function, and you don't—unless you've read my explanation that appears a bit later in this article.

Another important consideration to think about is that these worms are vindictive little bastards. When they die, they're going to give you the middle finger on their way out of your skin, brain, and/or colon. Specifically, the dying worms release toxins, and these can cause a full-body allergic reaction (the dreaded Jarisch-Herxheimer reaction). Especially at the beginning of therapy, you're going to feel really, really bad—like fetal-position-I-want-to-die-just-shut-up-and-let-me-sleep bad. (Consider this to be a happy sign that the medicine is working!) For me, I had to sleep within an hour of taking praziquantel. However, as my infectious load decreased, the overpowering urge to sleep diminished. Also, I would feel paralyzed temporarily upon waking, and I would have a painfully dry mouth (anti-cholinergic effects), also upon waking. Because of the way that the physical reactions diminished over time, I believe that they are due to the dying organisms rather than to the drug itself, but I could be wrong—it probably doesn't matter. The worst listed side effect of these drugs, praziquantel in particular, is vomiting, although I have almost never experienced that. More common are mild nausea, mild headache, and mild gastrointestinal cramps. Also, it makes food taste bad. (But it's not like the side effect is leukemia or aplastic anemia or sudden death or something.)

Praziquantel actually has fewer side effects than aspirin, and it appears to be much safer, too; it has been around since the mid-1970s. Praziquantel is on the World Health Organization's list of Essential Medicines, which are drugs that WHO thinks should be readily accessible to people at an affordable price in every country in the world (except the U.S., apparently— what about that, WHO?). The fact that it is virtually never prescribed in the U.S. demonstrates to me that U.S. doctors are out of touch with reality when it comes to worm infections. In Baltimore, Maryland, where I live, no pharmacies that I called carried either ivermectin or praziquantel, including the outpatient pharmacy at the world-famous Johns Hopkins Hospital, so that should tell you something. About 5 billion people in the world are infected with worms (at least!—*Current Medical Diagnosis and Treatment*), except in America, because we're too good for that?!! Come on, let's get real! Worms are here, and everyone has them!

Another useful factoid about praziquantel is that it acts against worms by blocking calcium channels, causing a build-up of calcium inside the worm that causes its paralysis. (I have read that some people no longer think that this might be true.) Therefore, it seems reasonable to take calcium supplements, which means that you also should take vitamin D supplements, and not the dry kind, because those are useless; vitamin D is a fat-soluble vitamin, meaning that you can't absorb it unless it is in an oil solution, so take the fish oil vitamin D gel capsules. And if you take calcium, you should also take magnesium, and sodium, and potassium, and copper, and zinc, and molybdenum, and it keeps getting even more complicated because they're all connected . . . .

Praziquantel and ivermectin are metabolized in the liver, although metabolites of praziquantel are eliminated in the urine (through the kidneys). Taking these drugs could be dangerous for people who have pre-existing liver problems (and possibly kidney problems). However, I have a <u>recipe to strengthen your liver:</u> First, you want to strengthen Phase 2 metabolism by taking N-acetylcysteine, taurine, and glycine. Then you want to boost Phase 1 metabolism by taking about 600 mg of alpha-lipoic acid, 900 mg of milk thistle or silymarin (the active ingredient in milk thistle), and 400 mcg (that's micrograms) of selenium per day for a few weeks. (I think that I got the second part of this recipe from Julian Whittaker MD's newsletter; the other part I developed myself. This recipe is supposed to be powerful enough to cure Hepatitis C, although I can't personally attest to that.) I do worry, however, that by boosting your own liver, you might also be assisting the worms in resisting the worm-killing drugs. Proceed with caution. This liver recipe should also help to treat hangovers, although because I regard drunkenness as stupid, I've never been drunk, so I can't really say. But just do not take acetaminophen (Tylenol) for hangovers, as I've been told some people do—that can kill you.

I hope you are now convinced that worms cause Parkinson's Disease/ MS, etc. I don't know if the method of treatment is especially important, but allow me to share a few observations on my personal treatment methodology. One thing that I did is to eat a small amount of my allergy-causing foods (wheat/gluten, milk/some dairy, mustard/salt/vinegar, etc.) immediately before taking my medicine. (This may or may not work with <u>peanut or seafood allergies</u>, but if you try to treat those using this method, be very, very careful, and do it in a hospital under a doctor's supervision, because, although my allergies were life-threatening, they were not as severe as peanut and seafood allergies tend to be, apparently.) My theory is that I wanted to catch the worms with their little worm mouths open in order to shove a whole bunch of toxic anti-worm medicine down their little worm throats. Did it make a difference? I can't say for sure, not having done one of those fancy-pants, double-blind, controlled studies, but I think it did. I can say, however, that my health has improved dramatically (I've lost more than 10 pounds of tumor-like lymph nodes) since I've started my anti-worm therapy, and I am having super-dramatically less pain, and I am bench-pressing almost more weight than ever. My therapy is not over, but I see the light at the end of the tunnel.

I have just been informed that some people are taking diatomaceous earth (DE) orally to kill worms. DE is a particular kind of sand that is used in swimming pool filters; it is also used as a non-toxic method of killing cockroaches. The theory here is that the sand lacerates the worms in the intestines. Also, smearing the DE around the outside of the rectum and genitals at bedtime prevents the worms from crawling out to reproduce (which normally causes rectal and genital itching at night—another sign of worm infection), which they need to do for some reason; there are several types of worms, usually roundworms, that perform this unusual procedure. DE is probably a good adjunctive treatment for worms in the intestines, but it will not treat worm infections that are disseminated throughout the body. Because it is just sand (silica), it probably won't hurt, but there may be a possibility of damaging the mucosa (lining) of the intestines, so I won't recommend taking it internally unreservedly at this point. But I just wanted to let you know about another possible adjunctive treatment, although it *is* good to apply to the rectum at bedtime. I have recently started applying DE to my rectum and genitals at nighttime, and it seems to help. In any event, it reduces the itching that is presumably caused by worms.

Important

Also, as the parasites/worms are killed off, I've noticed the onset of tremors, i.e. shaking hands, etc. In part this may be due to the fact that I now have to eat less often, and my body is being mildly starved as I lose weight for both intentional and unintentional reasons, which it has become unaccustomed to doing. One of the problems with starvation is something called "refeeding syndrome," which is caused by a deficiency of phosphorus/phosphate, which can be fatal in extreme cases. I have found that I can reduce my tremors by adding phosphorus/phosphate to my diet in the form of Baking Powder [the main ingredient of which should be monocalcium phosphate—I use Rumford™ brand] (not baking soda). (Use only aluminum-free baking powder—aluminum is poison!) It costs about $1.50 at the grocery store.

All you need to do is to put about ¼ teaspoon of baking powder into a glass of water and drink it. Or your doctors can put you on an intravenous drip of phosphate for about $1,000 or so—it's your choice! Don't forget to take your multi-mineral supplements! Milk is also high in phosphate, but it has calories and fat, which baking powder doesn't. It can also make the body too alkaline if you drink a lot of it.

Since writing the above, I've discovered a few additional findings. I seemed to reach a plateau in my treatment, or I just needed a break from the debilitating effects of killing off the parasitic organisms, so I took a break from treatment. Since, then, I've started up with the following regimen of anti-worm substances, which has produced very good results:

I've started drinking French-press <u>coffee</u>, with <u>cinnamon</u> brewed in.

I've started using more <u>curry powder/turmeric</u> in my food; curry powder often has <u>black pepper</u> in it, which contains piperine, which works synergistically with turmeric (curcumin), increasing the effectiveness many times. I also use extra black pepper.

I've been using a lot of <u>garlic</u> in my food, such as in hummus dip and spaghetti.

I've been using ¼ teaspoon or less of <u>boric acid</u> once or twice a week, shortly after taking my praziquantel.

I've been using <u>praziquantel</u>, dissolved in drinking alcohol occasionally one time per day before bed, on nights when I don't have to get up early the next day.

This regimen has produced accelerating shrinkage of my lymph nodes—a good thing. Interestingly, I just read that borates are often assayed (analytical quantification) by mixing them with curcumin (a main ingredient in turmeric), which forms some poorly-soluble chemicals called rosocyanine and rubrocurcumin. I don't know what these do, but apparently, no one else does, either, other than that they turn red, making them useful for measurement. Perhaps they are toxic to worms—but they could be toxic to people, too; or they may not be toxic at all. As the chief research monkey in my research laboratory, I can say that I'm not dead, yet, and I feel better so far, but I can't say more than that, at this point. [Note: this happens only at very high temperatures—not in living beings.]

Also interesting is that curcumin (accent on the first syllable), the main ingredient in the plant/spice turmeric (that's with two "R"s), a member of the ginger family, is made in plants from phenylalanine (an amino acid), which gets converted to cinnamic acid by deleting the nitrogen group. Cinnamic acid is a main ingredient in cinnamon (cinnamic acid is converted to cinnamaldehyde in cinnamon, or para-coumaric acid in turmeric to form curcumin), while phenylalanine is a precursor to tyrosine in humans, which is the usual starting point to make dopamine in humans and other animals. Cinnamic acid is then converted to para-coumaric acid before forming curcumin, para-coumaric acid being exactly the same as tyrosine, except for that aforementioned nitrogen group and an extra double bond that is required as a result of deleting the nitrogen group. This similarity leads me to suspect that some of the health effects of cinnamon and curcumin may come from their similarity to precursors of DOPA and dopamine, and that perhaps the analogs are blocking hormone synthesis in the parasites. Curcumin apparently does not penetrate the blood-brain barrier in humans, but I'll bet that cinnamic acid (or cinnamaldehyde, the main substance in cinnamon) and para-coumaric acid probably do! Perhaps cinnamon has some of the health effects of curcumin because they are so closely related. (Perhaps cinnamon has more health effects than does turmeric.) These ingredients of cinnamon and turmeric might inhibit the synthesis of hormones and proteins by preventing/blocking the molecules from forming the proper hormones and by terminating protein synthesis by forming only one instead of two peptide bonds. These effects would probably be stronger in parasites than in the human host, possibly because of higher metabolic rates in smaller organisms. In any event, the aforementioned nutritional substances (coffee, turmeric, black pepper, cinnamon, garlic, boric acid) seem to have infused new energy into my anthelminthic regimen, which makes me much happier. For a visual explanation of some of the above, see *Curcumin* on *Wikipedia*.

Another thing that I've been doing lately is to go down to my local sub sandwich shop and order a large submarine sandwich. I'll bring it home and load it up with curry powder, turmeric, salt, black pepper, cinnamon, and two large cloves of garlic. I'll eat the first half of the sandwich, and then I'll take ¼ teaspoon of boric acid along with my usual dose of praziquantel, and then I'll eat the second half of the sandwich. This seems to give a really big boost to my anti-worm regimen, and leaves me feeling hit very hard, especially the next morning, when I have to sleep late, until the effects of this powerful regimen wear off. It leads to increased shrinkage of my lymph nodes, which is perhaps the most objective measurement of my progress.

Also, I have recently started eating two slices of heavily-spiced cinnamon toast with butter and jam before bedtime, and I have noticed startling improvement as a result. Cinnamon may actually be almost as good as praziquantel in treating worms, which just seems amazing to me! This experiment continues...

Here is a list (in random order) of other foods that are reputed to have anthelminthic properties: curry leaves, pumpkin seeds and oil, savory, coconut, carrots, oregano, thyme, papaya (papain), pineapple, neem, cloves, radish, fenugreek (as a bonus, fenugreek is said to increase breast size in women), bitter gourd, mustard (has turmeric in it), celery, brassicas vegetables (e.g. broccoli, cauliflower, kale), licorice, oily fish, chamomile, chocolate, nuts, olive oil, berries, beets, pomegranate, cardamom, nutmeg, apple cider vinegar, cranberries, fennel seed tea, wormwood, black walnut, chilies, horseradish, cayenne, sauerkraut, rejuvelac, bee propolis, peppermint, caprylic acid, onions, eugenol, aloe, pomegranate, tansy, thuja, rue, tonka beans, lavender, sweet clover grass, apricots, strawberries, cherries, dong quai, and maybe: grape seed, pine bark extract, anise, noni, wintergreen, and ginger.

Note that just because something is listed here doesn't mean that it is safe or intended for internal use—for instance, rue is very toxic, hence the phrase, "you will rue the day that you crossed me," or something like that, which is generally regarded as a threat.

Finally, that old standby that has been used for more than 2,000 years, castor oil is quite probably very useful against worms, used internally or externally.

It is said to be very toxic when taken internally, but it is absorbed through the skin so easily that I doubt that there is really that much difference between the two routes of administration. I picked up a used book about castor oil at a library sale called, *The Oil That Heals*, by William A. McGarey, and I've found that castor oil seems to be beneficial—one more tool in the toolbox. There's too much information there to go into here, but feel free to get the book and investigate for yourself.

I hope that reviewing my experiences will help you and your doctor arrive at a therapeutic regimen that works well for you and helps you return to health.

As always, please vote for me for the Nobel Prize in Physiology or Medicine, if you are an eligible voter.

— Robert S Farmer, MD

REFERENCES

1   http://fnic.nal.usda.gov/dietary-guidance/dri-reports/vitamin-vitamin-k-arsenic-boron-chromium-copper-iodine-iron-manganese; *Copper*, USDA National Agricultural Library website: http://www.nal.usda.gov/fnic/DRI//DRI_Vitamin_A/224-257_150.pdf, accessed Nov. 17, 2012.

2   Turnlund JR. *Copper*. In: Shils M, Shike M, Ross AC, et al, eds. *Modern Nutrition in Health and Disease, 10th edition*. Lippincott Williams & Wilkins, Philadelphia, 2005.

www.ingramcontent.com/pod-product-compliance
Lightning Source LLC
Chambersburg PA
CBHW032004170526
45157CB00002B/536